D1555154

OXFORD MATHEMATICAL MONOGRAPHS

Editors

G. TEMPLE   G. HIGMAN

OXFORD MATHEMATICAL MONOGRAPHS

MEROMORPHIC FUNCTIONS

By W. K. HAYMAN. 1963

THE THEORY OF LAMINAR BOUNDARY LAYERS IN COMPRESSIBLE FLUIDS

By K. STEWARTSON. 1964

HOMOGRAPHIES, QUATERNIONS AND ROTATIONS

By P. DU VAL. 1964

CLASSICAL HARMONIC ANALYSIS AND LOCALLY COMPACT GROUPS

By H. REITER. 1968

QUANTUM-STATISTICAL FOUNDATIONS OF CHEMICAL KINETICS

By S. GOLDEN. 1969

VARIATIONAL PRINCIPLES IN HEAT TRANSFER

By MAURICE A. BIOT. 1970

PARTIAL WAVE AMPLITUDES AND RESONANCE POLES

BY

J. HAMILTON

AND

B. TROMBORG

OXFORD

AT THE CLARENDON PRESS

1972

Oxford University Press, Ely House, London W. 1

GLASGOW NEW YORK TORONTO MELBOURNE WELLINGTON
CAPE TOWN IBADAN NAIROBI DAR ES SALAAM LUSAKA ADDIS ABABA
DELHI BOMBAY CALCUTTA MADRAS KARACHI LAHORE DACCA
KUALA LUMPUR SINGAPORE HONG KONG TOKYO

PRINTED IN GREAT BRITAIN
AT THE UNIVERSITY PRESS, OXFORD
BY VIVIAN RIDLER
PRINTER TO THE UNIVERSITY

PREFACE

THIS monograph grew from an article which we intended to write in order to clear our own minds on some of the more mathematical aspects of the partial wave amplitudes used for the binary collisions of elementary particles. This origin explains why we have not attempted to give a complete treatment of partial wave amplitudes. For physicists, we can say that we have assumed about the average knowledge of partial wave amplitudes which an elementary particle theorist would have. On the other hand, we feel that a mathematician might well consider that the book is self-contained, at least as far as the elementary material is concerned.

We would like to thank our colleagues Jens Lyng Petersen and Gösta Gustafson for valuable help.

J. H.
B. T.

Copenhagen
February 1972

CONTENTS

1

INTRODUCTION

1.1. The physical background

We shall first give a brief account of the physical background of the mathematical problems that are to be discussed. These problems arise in the theory of the strong interactions of elementary particles, and it is necessary to discuss the nature of this theory since it differs considerably in outlook and in method from earlier physical theories.

The strong interactions of elementary particles occur at or within a range of approximately 10^{-13} cm; the size of the strongly interacting particles is also of that order. The interactions can be due to the exchange of one or more elementary particles or resonances; the interactions are further complicated by the ease with which other particles can be created. The methods of relativistic quantum field-theory will describe certain general features but they cannot in general be used for detailed calculations. This limitation is associated with the complicated nature of any field description of strongly interacting elementary particles; we have to look on a particle as in some sense comprising an assortment of other particles, all of them being strongly interacting.

Fortunately we can retain a considerable degree of simplicity by using the S-matrix description of the collision processes between elementary particles. This relates the asymptotic outgoing states in a collision to the asymptotic incoming states. It has been discovered how to use and develop the S-matrix method so as to provide a physical description and understanding of numerous elementary particle interactions and of the structure of various particles; a means of making predictions in a variety of situations has also been developed.

The S-matrix technique is based on the three physical principles of causality, crossing, and unitarity. (Further details of these three principles may be found in the works quoted in Hamilton [29].) *Causality* requires that the scattering amplitudes (i.e. the elements of the S-matrix) are analytic functions of the kinematic variables of the incoming and outgoing particles. We shall be concerned with two-body reactions where there are two incoming and two outgoing particles. The scattering amplitudes $A(s, t)$ are analytic functions of s and t, the squares of the energy and of the momentum transfer in the centre of mass system (c.m.s.). (When the particles are spinless we have only one amplitude $A(s,t)$ describing the reaction.) These functions are regular in the two variables s and t except for certain singularities whose properties and locations are related to fairly simple physical properties of the system. For example the amplitudes for elastic nucleon–nucleon scattering (NN $\rightarrow$ NN) can have simple poles in the energy corresponding to the fact that the deuteron is an NN bound state. They have a cut along that portion of the real energy axis for which scattering can occur (this is called the physical cut). They also have a simple pole in t arising from the fact that a pion (π) can be exchanged between the two nucleons. They also have a cut in t arising from the possibility of exchanging two or more mesons (of any physical total energy) between the two nucleons. The residue of the pion pole is given by the πNN coupling constant, and the discontinuity across the cut arising from the exchange, for example, of two pions measures the strength of the two-pion exchange interaction. Other singularities of the amplitudes follow from crossing.

Crossing is a consequence of general features of relativistic quantum field-theory. It relates particle processes to anti-particle processes. For example in pion–nucleon scattering (πN $\rightarrow$ πN), crossing tells us that the amplitudes for πN $\rightarrow$ πN are the same analytic functions as the amplitudes for $\pi\bar{\text{N}} \rightarrow \pi\bar{\text{N}}$ and as the amplitudes for $\pi\pi \rightarrow \text{N}\bar{\text{N}}$, where $\bar{\text{N}}$ denotes an anti-nucleon. All we need do to get from one process to another is to continue analytically in the energy or momentum transfer or both. Crossing is essential in specifying the interactions. An important interaction in πN $\rightarrow$ πN is nucleon exchange; crossing relates this directly to the process πN $\rightarrow$ N $\rightarrow$ πN which is described by a simple pole (the nucleon pole).

Unitarity relates the imaginary (i.e. the absorptive) part of a scattering amplitude to an integral over a bilinear product of scattering amplitudes. The optical theorem, which expresses the imaginary part of the forward amplitude in terms of the total cross-section, is a special case. Unitarity imposes an important restriction on the discontinuity of the analytic functions across the physical cut. Moreover it provides further information about the interactions; for example, the two-pion exchange interaction in πN $\rightarrow$ πN can now be related to pion–pion scattering ($\pi\pi \rightarrow \pi\pi$).

1.2. Partial wave amplitudes

On specifying the angular momentum in a collision process we obtain partial wave amplitudes (p.w.a.). These are readily derived from the scattering amplitudes, and from a physical viewpoint have considerable simplifying features. Bound states and resonances occur in individual p.w.a. and advantages can be derived from the presence of the centrifugal force. For two-body collisions the p.w.a. have simple analytic properties, and the unitarity relation has a simple form.

This work deals with aspects of the theory of partial wave amplitudes for two-body collisions, and in particular with the p.w.a. leading to final particles which are identical with the initial particles. If the particles are spinless, we define partial wave projections $A_l(s)$ $(l = 0, 1, 2,...)$ by the equation

$$A_l(s) = \tfrac{1}{2}\int_{-1}^{1} A(s,t)P_l(\cos\theta)\,\mathrm{d}(\cos\theta),$$

where l is the orbital angular momentum, P_l is the Legendre polynomial, and θ is the scattering angle. This relation gives the singularity structure of $A_l(s)$. It is regular in the s-plane except for a physical cut $s_0 \leqslant s \leqslant \infty$, a so-called left-hand cut, and possibly bound state poles. The left-hand cut of $A_l(s)$, although generally not the whole left-hand cut, is expressed simply in terms of the exchange interactions which produce the driving forces. However, the situation is more complicated for particles with spin. The partial wave projections $A_l(s)$ are simply related to the usual p.w.a. $f_l(s)$.

1.3. Survey of the problems

Uniqueness

Using a Cauchy integral we can express a p.w.a. $f_l(s)$ (or a partial wave projection $A_l(s)$) in terms of its singularities in the s-plane; such an equation is called a *dispersion relation*. The problem is whether the p.w.a. is uniquely determined by the discontinuity across the left-hand cut and by the unitarity relation on the physical cut.

The problem of the uniqueness of solutions of partial wave dispersion relations (p.w.d.r.) has been known for a long time in connection with CDD (Castillejo–Dalitz–Dyson) poles [1], and it is usually discussed in terms of the N/D method, as, for example, in [2] and [3]. Such treatments are apt at least to look complicated, and we shall here give a simple and direct approach to the uniqueness problem. A similar discussion may be found in [4].

We must first comment on the description of any inelasticity that may be present. The p.w.a. $f_l(s)$ is normalized so that on the physical cut

$s_0 \leqslant s \leqslant \infty$, writing $s+$ for $s+i\epsilon$ where ϵ is small and positive,

$$f_l(s+) = \frac{\eta(s)\exp(2i\alpha(s))-1}{2iq}. \tag{1.1}$$

Here q is the momentum in the c.m.s., $\alpha(s)$ is the real part of the phase shift, and $\eta(s)$ (where $0 \leqslant \eta \leqslant 1$) is the inelasticity coefficient. For a purely elastic reaction $\eta = 1$. However, scattering is not elastic in general, and we shall assume that $\eta(s)$ is a given function in the physical region $s_0 \leqslant s \leqslant \infty$. Then we can use Froissart's transformation [5] to obtain an equivalent elastic amplitude.

It should be noted that from a physical viewpoint it may not always be suitable to prescribe $\eta(s)$. For example, near a resonance it might be better to describe the inelasticity by $R = \sigma_{\text{tot}}(s)/\sigma_{\text{el}}(s)$ where σ_{tot} and σ_{el} are the total and the elastic cross-sections for the partial wave. However, in order to avoid complications we shall assume $\eta(s)$ is given.

We shall discuss the uniqueness problem for two types of p.w.d.r. These types are: (*a*) p.w.d.r. for $A_l(s)$ in which we specify the left-hand discontinuity of $A_l(s)$; (*b*) p.w.d.r. for $f_l(s)$ in which the discontinuity of $f_l(s)$ across the left-hand cut is specified. For the scattering of *equal mass particles*,

$$A_l(s) = s^{\frac{1}{2}} f_l(s),$$

where $s^{\frac{1}{2}}$ is defined as in Fig. 1.1.

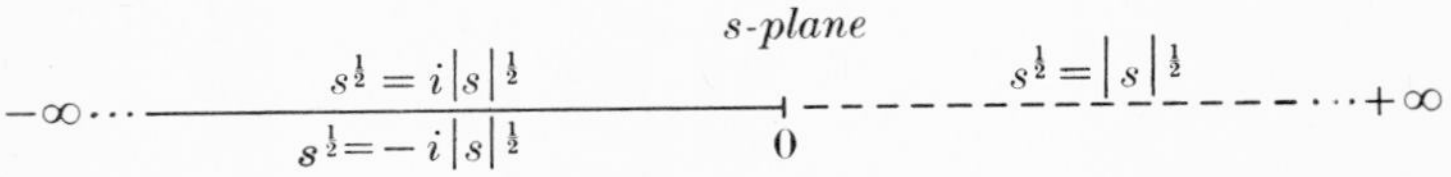

s-plane

$-\infty \cdots$ —— $s^{\frac{1}{2}} = i|s|^{\frac{1}{2}}$ / $s^{\frac{1}{2}} = -i|s|^{\frac{1}{2}}$ —— 0 - - - $s^{\frac{1}{2}} = |s|^{\frac{1}{2}}$ - - - $\cdots +\infty$

FIG. 1.1. Definition of $s^{\frac{1}{2}}$ in the s-plane cut along $-\infty \leqslant s \leqslant 0$.

In equal mass scattering the left-hand cut is simply a segment $-\infty \leqslant s \leqslant s_1$ of the real axis, and the left-hand cut discontinuities of A_l and f_l are given by $\operatorname{im} A_l(s+)$ and $\operatorname{im} f_l(s+)$ $(-\infty \leqslant s \leqslant s_1)$. Since

$$\operatorname{im} A_l(s+) = |s|^{\frac{1}{2}} \operatorname{im} f_l(s+), \quad \text{for } s > 0,$$

and
$$\operatorname{im} A_l(s+) = |s|^{\frac{1}{2}} \operatorname{re} f_l(s+), \quad \text{for } -\infty \leqslant s < 0,$$

we see that in the case of equal mass heavy particle (e.m.h.) scattering, e.g. NN-scattering, where s_1 is large and positive, the discontinuity of A_l gives $\operatorname{im} f_l(s+)$ on a large segment $(0 < s \leqslant s_1)$ of the left-hand cut. In such a case the p.w.d.r. of type (*b*) is also of practical importance.

Partial wave dispersion relations of type (*b*) can also be relevant in some cases of unequal mass scattering, as for example in some treatments of πN scattering. An example is given in Chapter 6. In $\pi\pi$-

scattering the left-hand cut is the segment $-\infty \leqslant s \leqslant 0$, so here it is only p.w.d.r. of type (a) which are relevant in practice.

The main result of the uniqueness problem (Chapter 2) is that there is at most one isolated solution—the discrete solution. Other solutions are member of a manifold of solutions. The condition for the isolated solution and the dimension of any manifold of solutions are simply expressed in terms of the phase shift at infinite energy $\alpha(\infty)$ and the behaviour of $\eta(s)$ as $s \to \infty$. There is a close relationship with Levinson's theorem. In Chapter 2 we also give examples which illustrate the CDD pole aspect.

Second sheet poles

Unitarity shows that there is a square root branch point at the physical threshold s_0. The unitarity relation makes it possible to continue the p.w.a. (or the p.w. projection) through the physical cut on to what we call the *second sheet*. Cuts on the second sheet can only lie in the same positions as cuts on the first sheet. However, the p.w.a. can have poles on the second sheet, and a second sheet pole which lies close to the physical cut represents a resonance. It is important for the physicist to know where the second sheet poles can lie, how their positions depend on the strength of the interaction, and whether the correlation of poles with resonances persists as the poles move far away from the physical cut. We shall discuss these questions for elastic scattering, but the Froissart transformation can be used to relate the results to inelastic cases.

For elastic scattering the function

$$S(s) = 1+2iqf_l(s)$$

has unit modulus on the physical cut. In Chapter 3 we study the unimodular curves $|S(s)| = 1$ on which a real phase shift can be defined. These, together with the level curves $\operatorname{re} S(s) = 0$ and $\operatorname{im} S(s) = 0$, provide useful restrictions on the location of second sheet poles.

Resonance can arise naturally in P-waves when the interaction is attractive, and we devote several chapters to deriving theorems on the location of the second sheet poles for such amplitudes, also deriving theorems on the motion of the poles as the strength of the attractive interaction is varied. In this work only the isolated solutions are considered. Inelastic as well as elastic amplitudes are discussed.

In Chapter 4 we shall deal with equal mass heavy particles (case (b) above) in an attractive interaction of bounded range. The results include the proof that there cannot be more than one pair of second sheet poles away from the real axis, and the proof that the second sheet poles must lie in a region re $s \leqslant c$ where the finite number c is determined by the range of the interaction. Suppose the resonance position s_R is determined by the condition $\alpha(s_R) = \frac{1}{2}\pi$. Then it can be shown that there exists a band of values of interaction strength such that s_R is a large distance away from the second sheet poles. However, when s_R is not far from the threshold s_0, it will lie near the resonance poles.

In Chapter 5 the analogous properties are investigated for the attractive P-wave π–π case (case (a) above), again using the isolated solution. There are various important differences in the theorems which can be proved. The unequal mass case, such as in πN scattering, is in general a complicated problem, and in Chapter 6 we show a special example of an attractive πN P-wave which is a rough approximation to the N_{33} resonance; various interesting properties of that example are given.

It should be emphasized that in the work on P-wave amplitudes in Chapters 4–6 we shall only consider left-hand cuts which contain no repulsive elements. It is expected that many of our results will not be valid if part of the interaction is repulsive, even though the over-all effect is an attraction. The theorems which could be derived in that situation are not likely to be so simple as our results.

Finally in Chapter 7 some *pathological resonances* are discussed. These are partial wave amplitudes whose phase shift passes up through $\frac{1}{2}\pi$ as s increases; in this case there are no second sheet poles lying off the real s-axis. Various types are known; some contain essential singularities and are presumably unrelated to physics, but other examples do not have this defect. Classes of such amplitudes and their properties are examined. We have not discussed the decay law for such resonances.

2

ARBITRARINESS OF SOLUTIONS OF PARTIAL WAVE DISPERSION RELATIONS

2.1. Specifying $\operatorname{im} f(s)$ on the left-hand cut

THE partial wave amplitude (p.w.a) $f(s)$ has the physical cut $s_0 \leqslant s \leqslant \infty$ and the unphysical (or left-hand) cut $-\infty \leqslant s \leqslant s_1$ where $s_1 < s_0$ (Fig. 2.1). It obeys the reality condition

$$f^*(s) = f(s^*)$$

for all values of s on the physical sheet, and on the physical cut it takes the form shown in eqn (1.1). The discontinuity across the left-hand cut is

$$2i \operatorname{im} f(s+) \equiv 2i\rho(s), \quad \text{for } -\infty \leqslant s \leqslant s_1.$$

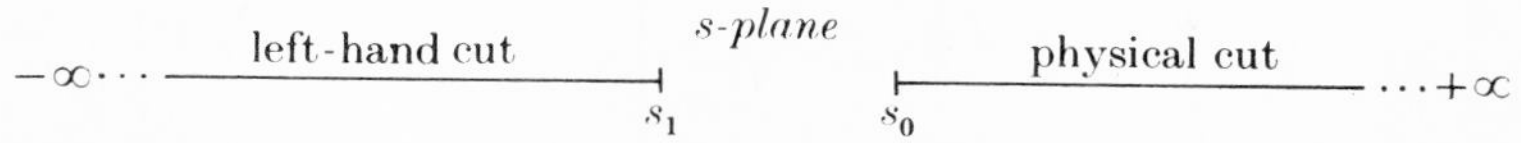

$-\infty \cdots$ left-hand cut ⊣ s_1 *s-plane* s_0 ⊢ physical cut $\cdots +\infty$

FIG. 2.1. The physical cut $s_0 \leqslant s \leqslant \infty$ and the unphysical cut (left-hand cut) $-\infty \leqslant s \leqslant s_1$ of a typical partial wave amplitude $f(s)$.

As we discussed in Chapter 1, we can cover a number of important cases (type (*b*)) by assuming that on $-\infty \leqslant s \leqslant s_1$, $\operatorname{im} f(s+)$ is given.

The p.w.d.r is either

$$f(s) = \frac{1}{\pi}\int\limits_{-\infty}^{s_1} \frac{\rho(s')\,ds'}{s'-s} + \frac{1}{\pi}\int\limits_{s_0}^{\infty} \frac{\operatorname{im} f(s'+)\,ds'}{s'-s}, \tag{2.1}$$

or the equivalent relation with a finite number of subtractions. (In § 2.5 it is shown that, for a large class of p.w.a. $f(s)$, no subtraction is required. We shall almost entirely confine ourselves to this class.)

Given $\rho(s)$ on $-\infty \leqslant s \leqslant s_1$, we have to find a solution $f(s)$ of eqn (2.1) which obeys the unitarity relation on the physical cut,

$$|2qf(s+)-i| = \eta(s), \quad \text{for } s_0 \leqslant s \leqslant \infty. \tag{2.2}$$

Here $\eta(s)$ is given and $0 < \eta(s) \leqslant 1$ $(s_0 \leqslant s < \infty)$ (we do not allow $\eta(s) = 0$, except when $s \to +\infty$). The important question is whether this solution $f(s)$ is unique.

For most of this chapter we shall discuss this uniqueness problem (type (*b*)), and we should remark that it is not necessary that all the unphysical cuts lie on the real axis. For unphysical cuts off the real axis, in place of $\rho(s)$ we must specify $(1/2i)\Delta f(s)$ where $\Delta f(s)$ is the discontinuity in $f(s)$ across the cut. Also it is not necessary to specify the kinematical relation $q^2(s)$, except to say that $q^2 \to (\text{constant}).s$ as $|s| \to \infty$, and that $q(s)$ has a square root branch point at the threshold s_0.

First we shall use a simple method and later (in §§ 2.3 and 2.4) we shall give a better and more powerful method; a similar method has been given by Frye and Warnock [2], [3] and by Lyth [4]. In § 2.7 we shall examine the uniqueness problem for type (a), i.e. when $\operatorname{im} A_l(s)$ is specified on the left-hand cut.

Various related topics such as the behaviour at infinity, and the relation to CDD poles, are discussed in §§ 2.5 and 2.6.

The simple method: no bound state

Suppose we have a solution $f(s)$ obeying eqns (2.1) and (2.2), and let $\alpha(s)$ be the real part of its phase on the physical cut. Let there be another solution $(f(s)+\delta f(s))$ whose phase is $(\alpha(s)+\delta\alpha(s))$ where $\delta\alpha(s)$ is small $(s_0 \leqslant s \leqslant \infty)$. By eqn (1.1)

$$\delta f(s+) = \frac{\eta(s)\delta\alpha(s)}{q}\exp(2i\alpha(s)), \quad \text{for } s_0 \leqslant s \leqslant \infty, \tag{2.3}$$

since $\eta(s)$ is fixed. Consequently

$$\delta f(s+) = \pm\exp(2i\alpha(s))|\delta f(s+)|, \quad \text{for } s_0 \leqslant s \leqslant \infty, \tag{2.4}$$

where the sign reverses at any simple zero of $\delta f(s)$.

The analytic function $\delta f(s)$ has a cut $s_0 \leqslant s \leqslant \infty$ on which it obeys eqn (2.4). Since $\rho(s)$ is not varied, $\delta f(s)$ has no cut along $-\infty \leqslant s \leqslant s_1$. Here we assume there is no bound state, therefore $\delta f(s)$ has no other singularity on the physical sheet.

Using a method similar to that discussed by Lovelace [6], in the context of symmetry breaking, we shall consider the function

$$\Lambda(s) = \exp\left\{\frac{2s}{\pi}\int_{s_0}^{\infty}\frac{\alpha(s')\,ds'}{s'(s'-s)}\right\}. \tag{2.5}$$

The only singularity of $\Lambda(s)$ is the cut $s_0 \leqslant s \leqslant \infty$ and on this cut

$$\Lambda(s\pm) = |\Lambda(s)|\exp(\pm 2i\alpha(s)), \quad \text{for } s_0 \leqslant s \leqslant \infty, \tag{2.6}$$

and

$$|\Lambda(s)| = \exp\left\{\frac{2s}{\pi}P\int_{s_0}^{\infty}\frac{\alpha(s')\,ds'}{s'(s'-s)}\right\} \tag{2.7}$$

(where $s\pm = s\pm i\epsilon$). The choice $\alpha(s_0) = 0$ is necessary to ensure $|\Lambda(s)|$ is finite at the threshold s_0.

If δc is a small real number, then

$$\delta f(s) = \delta c \,.\, \Lambda(s) \tag{2.8}$$

obeys eqn (2.4), and has no singularity apart from the cut $s_0 \leqslant s \leqslant \infty$. Strictly speaking, to avoid errors of order $(\delta\alpha(s))^2$ in eqn (2.8) we should use $\delta f(s)/\delta c = \Lambda(s)$ in place of eqn (2.8). A more general solution is

$$\delta f(s) = E(s)\Lambda(s), \tag{2.9}$$

where $E(s)$ is an entire function. Restricting the p.w.a. to those which obey dispersion relations with a finite number of subtractions, $E(s)$ has to be a polynomial in s. Its coefficients are infinitesimal real numbers.

Behaviour as $s \to +\infty$

The possible solutions are further reduced by looking at the behaviour as $s \to +\infty$. In order to deal with $\Lambda(s)$ we assume that $\alpha(s)$ obeys two conditions. The first is the Lipschitz condition,

$$|\alpha(s)-\alpha(s')| < K_1|s-s'|^{\nu}, \quad \text{for } s,\, s' \in [s_0, \infty], \tag{2.10}$$

where K_1 and ν are positive constants. Second, that

$$\lim_{s\to\infty} \alpha(s) = \alpha(\infty)$$

exists, and

$$|\alpha(s)-\alpha(\infty)| < K_2/s^{\beta}, \tag{2.11}$$

where K_2 and β are positive constants.

It is shown in Appendix I (Theorem A) that conditions (2.10) and (2.11) are sufficient to ensure that

$$|\Lambda(s)| \to Cs^{-2\alpha(\infty)/\pi}, \quad \text{as } s \to +\infty, \tag{2.12}$$

where C is a constant. The proof in Appendix I does not require so strong a convergence condition as eqn (2.11); however, eqn (2.11) is the natural form to use in connection with the amplitude $f(s)$, and this condition will be used in §§ 2.3 and 2.4 below. Notice also that conditions (2.10) and (2.11) are somewhat weaker than the condition

$$|\alpha(s)-\alpha(s')| < K\left|\frac{1}{s}-\frac{1}{s'}\right|^{\beta}, \quad \text{for } s,\, s' \in [s_0, \infty], \tag{2.13}$$

which is sometimes used (see, for example, Tricomi [7], Chap. 4, § 3, or Hamilton and Woolcock [8], pp. 745–6) to derive eqn (2.12).

Since $\delta f(s)$ is the difference of two solutions which are of the form in eqn (1.1), it follows that

$$|\delta f(s)| < K'\eta(s)s^{-\frac{1}{2}}, \quad \text{as } s \to +\infty,$$

where K' is a constant. If the largest value of λ ($\lambda \geqslant 0$) is chosen such that

$$s^{\lambda}\eta(s) < B, \quad \text{as } s \to +\infty,$$

where B is any constant, then the boundary condition for $\delta f(s)$ is

$$|\delta f(s)| < K''.s^{-(\frac{1}{2}+\lambda)}, \quad \text{as } s \to +\infty, \tag{2.14}$$

where K'' is a constant.

Discreteness condition

If the partial wave has orbital angular momentum l ($l > 0$) then $f(s)$ must vanish like $(s-s_0)^l$ near the physical threshold s_0. The same holds for $\delta f(s)$, and since $\Lambda(s_0) \neq 0$, the polynomial $E(s)$ of eqn (2.9) must contain the factor $(s-s_0)^l$. We write

$$E(s) = (s-s_0)^l\bar{E}(s), \tag{2.15}$$

where $\bar{E}(s)$ is a real polynomial of degree $(p-1)$ ($p \geqslant 1$).

Using eqns (2.9) and (2.12), the condition (2.14) requires

$$p \leqslant \frac{2\alpha(\infty)}{\pi} - \lambda - l + \tfrac{1}{2}. \tag{2.16}$$

(However, the subtler method in § 2.3 shows that there should be a strict inequality in this equation.) If no *positive* integer p obeys eqn (2.16) the function $\delta f(s)$ cannot exist and the original solution $f(s)$ is isolated. If positive integers p can satisfy the equation, the largest such integer gives the degree of arbitrariness of the solution of the p.w.d.r., since it gives the number of arbitrary infinitesimal real parameters in $\bar{E}(s)$. We call this number the *index*.

If the solution $f(s)$ is not isolated, we can build up a finite deviation $\Delta f(s)$ from $f(s)$ by successive infinitesimal deviations $\delta f(s)$:

$$\Delta f(s) = \int \delta f(s).$$

The integration is over the infinitesimal parameters appearing in $\bar{E}(s)$. We then get a p-dimensional manifold of solutions $f(s)+\Delta f(s)$.

For example if there is a P-wave without a bound state, having $\alpha(\infty) = \pi$ and $\eta(s) \to \eta_\infty$ as $s \to +\infty$, η_∞ being a positive constant, then $\lambda = 0$, and the only positive integer obeying eqn (2.16) is $p = 1$. There is one degree of arbitrariness and $f(s)+\delta f(s)$ is a solution, with

$$\delta f(s) = \delta c(s-s_0)\Lambda(s), \tag{2.17}$$

δc being an infinitesimal real number. On the other hand if $\lambda > \frac{1}{2}$ the P-wave solution having $\alpha(\infty) = \pi$ is isolated.

For a given value of $\alpha(\infty)$ each unit increase in the orbital angular momentum l reduces the degree of arbitrariness by unity. This is because of the stronger centrifugal force restriction near s_0. The same happens when λ is increased by unity. This is not surprising since the larger value of λ will in effect impose a new sum rule on $f(s)$ on account of the more rapid decrease in $\operatorname{re} f(s)$ as $s \to +\infty$ which is required.

We shall in the applications mainly be concerned with the isolated solution ($p < 1$). It may, however, be of interest to notice that in the above case of the elastic P-state having $\alpha(\infty) = \pi$ and no bound state, for which $p = 1$, the arbitrariness consists chiefly in the possibility of increasing or decreasing the energy of the resonance. The width and shape of the resonance only change slowly. These results can be demonstrated by evaluating the function $\Lambda(s)$ for such a resonance.

2.2. Simple method: one bound state

Let the single bound state pole in $f(s)$ be

$$\frac{\Gamma}{s-B}, \quad \text{for } -\infty < B < s_0.$$

If it is specified that Γ and B are kept fixed, then $\delta f(s)$ has no singularity apart from the physical cut $s_0 \leqslant s \leqslant \infty$, and the analysis and results are identical with the case in which there are no bound states.

However, in general it is desired to determine the bound-state parameters B and Γ from the exchange forces. So we should inquire to what extent B, Γ, and $\alpha(s)$ ($s_0 \leqslant s \leqslant \infty$) are determined by the discontinuity $\rho(s)$ on the left-hand cut ($-\infty \leqslant s \leqslant s_1$) (compare with eqn (2.1)) plus the unitarity relation (eqn (2.2)).

When B and Γ are allowed to vary, $\delta f(s)$ will contain the pole terms

$$\frac{\delta\Gamma}{s-B}+\frac{\Gamma\,\delta B}{(s-B)^2}, \tag{2.18}$$

δB and $\delta\Gamma$ being the small changes in B and Γ. The only other singularity of $\delta f(s)$ is the cut $s_0 \leqslant s \leqslant \infty$, which must obey eqn (2.4). The general form of $\delta f(s)$ obeying these restrictions is

$$\delta f(s) = \left[\frac{\delta\Gamma}{s-B}-\frac{\Gamma\,\delta B}{s-B}\frac{\mathrm{d}}{\mathrm{d}B}\ln\{\Lambda(B)(B-s_0)^l\}+\frac{\Gamma\,\delta B}{(s-B)^2}+\bar{E}(s)\right]\times$$
$$\times\frac{(s-s_0)^l}{(B-s_0)^l}\frac{\Lambda(s)}{\Lambda(B)} \tag{2.19}$$

where $\bar{E}(s)$ is a real polynomial of degree $(p-1)$ ($p \geqslant 1$). The purpose of the second term in the square brackets is to ensure that $\delta f(s)$ has the form in eqn (2.18) near $s = B$ when $\delta B \neq 0$.

From the behaviour of eqn (2.19) as $s \to +\infty$ it can be deduced that:

(*a*) if†
$$0 \leqslant \frac{2\alpha(\infty)}{\pi} - l - \lambda + \tfrac{1}{2} < 1$$

there cannot be a polynomial $\bar{E}(s)$ and the only arbitrariness in the solution $f(s)$ arises from the changes δB and $\delta \Gamma$ in the bound-state parameters;

(*b*) if
$$-1 \leqslant \frac{2\alpha(\infty)}{\pi} - l - \lambda + \tfrac{1}{2} < 0,$$

we can only have

$$\delta f(s) = \frac{\Gamma\,\delta B}{(s-B)^2}\frac{(s-s_0)^l}{(B-s_0)^l}\frac{\Lambda(s)}{\Lambda(B)},$$

and because the term in $(s-B)^{-1}$ cannot occur, it follows that the variation δB in this case induces a variation $\delta\Gamma$ in the residue where

$$\frac{\delta\Gamma}{\Gamma} = \delta \ln\{\Lambda(B)(B-s_0)^l\};$$

(*c*) if
$$\frac{2\alpha(\infty)}{\pi} - l - \lambda + \tfrac{1}{2} < -1,$$

there is no arbitrariness in the solution and Γ and B are determined as well as $\alpha(s)$.

For example if we try to determine the nucleon as the bound state in a p.w.d.r. for the P_{11} π–N amplitude, the complete determination requires (case (*c*))
$$\alpha(\infty) < \tfrac{1}{2}\pi(\lambda - \tfrac{1}{2}).$$
However, if we assume the mass (i.e. $B = M^2$), then in order to determine the π–N coupling constant G^2 (i.e. Γ) we require (case (*b*))
$$\alpha(\infty) < \tfrac{1}{2}\pi(\lambda + \tfrac{1}{2}).$$
The P_{11} phase shift $\alpha(s)$ is uniquely determined (case (*a*)) for
$$\alpha(\infty) < \tfrac{1}{2}\pi(\lambda + \tfrac{3}{2}),$$
provided B and Γ are given.

For an elastic S-state, determination of the (single) bound state requires $\alpha(\infty) < -\tfrac{3}{4}\pi$. It is interesting to notice that Levinson's theorem, discussed in § 2.4, in this case gives $\alpha(\infty) = -\pi$ (we have chosen $\alpha(s_0) = 0$).

2.3. Improved method: no bound state

The improved method which we now present is much more precise and it enables us to derive theorems about the isolated solution or the

† In the better treatment (§ 2.4) the $\leqslant$ and $<$ signs in cases (*a*) and (*b*) are exchanged, and $<$ is replaced by $\leqslant$ in case (*c*) (see eqn 2.48).

dimension of the manifolds of solutions. Furthermore, given one solution we can find the explicit form of any other solution. G. Frye and R. L. Warnock [2], [3] and D. H. Lyth [4] have used a similar method and they have found results which to some extent overlap the results which we shall derive. The article by D. H. Lyth appeared while we were preparing the final version of this chapter. We shall not refer in detail to the interesting articles [2], [3], [4], partly because the notation is different, and partly because our method fits in more smoothly with the other topics which we shall discuss.

Froissart's transformation

We shall first introduce Froissart's transformation [5] to obtain from $f(s)$ the equivalent elastic amplitude $\tilde{f}(s)$. The function $(s-s_0)^{\frac{1}{2}}$ is defined

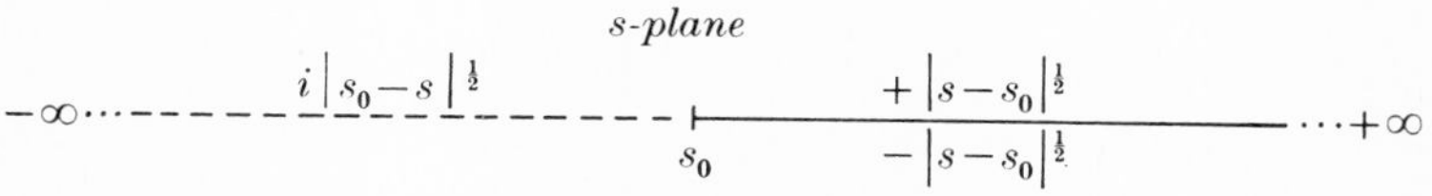

s-plane

$-\infty \cdots$ $i|s_0-s|^{\frac{1}{2}}$ s_0 $+|s-s_0|^{\frac{1}{2}}$ $-|s-s_0|^{\frac{1}{2}}$ $\cdots+\infty$

FIG. 2.2. The function $(s-s_0)^{\frac{1}{2}}$ defined in the s-plane cut along $s_0 \leqslant s \leqslant \infty$.

in the plane cut along $s_0 \leqslant s \leqslant \infty$ as shown in Fig. 2.2. On $-\infty \leqslant s \leqslant s_0$ its value is $i(s_0-s)^{\frac{1}{2}}$. The function

$$L(s) = \exp\left\{\frac{i(s-s_0)^{\frac{1}{2}}}{\pi}\int\limits_{s_0}^{\infty} ds' \frac{\ln \eta(s')}{(s'-s_0)^{\frac{1}{2}}(s'-s)}\right\} \tag{2.20}$$

is a regular function of s except for the cut $s_0 \leqslant s \leqslant \infty$ and a singularity at $s = \infty$. We assume that $\ln \eta(s_0)$ is either zero or finite, so near s_0

$$L(s) = 1+O((s-s_0)^{\frac{1}{2}}).$$

For our method it is also necessary to assume that $\eta(s)$ does not vanish quicker than a power of s as $s \to +\infty$.

On the cut $s_0 \leqslant s \leqslant \infty$,

$$L(s\pm) = \frac{1}{\eta(s)}\exp(\pm i\phi(s)), \tag{2.21}$$

where
$$\phi(s) = \frac{(s-s_0)^{\frac{1}{2}}}{\pi} P\int\limits_{s_0}^{\infty} ds' \frac{\ln \eta(s')}{(s'-s_0)^{\frac{1}{2}}(s'-s)}. \tag{2.22}$$

The equivalent elastic amplitude $\tilde{f}(s)$ is defined by

$$2iq(s)\tilde{f}(s)+1 = L(s)\{2iq(s)f(s)+1\}. \tag{2.23}$$

By eqn (2.21), on the physical cut,

$$\tilde{f}(s+) = \frac{\exp(2i\tilde{\alpha}(s))-1}{2iq}, \quad \text{for } s_0 \leqslant s \leqslant \infty, \tag{2.24}$$

and the equivalent elastic phase $\tilde{\alpha}(s)$ is

$$\tilde{\alpha}(s) = \alpha(s)+\tfrac{1}{2}\phi(s). \tag{2.25}$$

Near s_0, $\phi(s) \sim (s-s_0)^{\frac{1}{2}}$, so for $l \geqslant 1$, $\tilde{\alpha}(s)$ behaves like $(s-s_0)^{\frac{1}{2}}$ whereas $\alpha(s)$ behaves like $(s-s_0)^{l+\frac{1}{2}}$.

Suppose that

$$\eta(s) \to As^{-\lambda}\{1+Bs^{-(\frac{1}{2}+\epsilon)}\}, \quad \text{as } s \to +\infty, \tag{2.26}$$

where A and B are constants, and the constants λ and ϵ obey $\lambda > 0$ and $\epsilon > 0$. By eqn (2.22), for large positive s,

$$\phi(s) = \frac{s^{\frac{1}{2}}}{\pi} P \int_0^\infty \mathrm{d}s' \frac{\ln A - \lambda \ln s'}{(s')^{\frac{1}{2}}(s'-s)} + O(s^{-\frac{1}{2}}).$$

Using eqns (I.4) and (I.6) of Appendix I, this gives

$$\phi(s) = -\lambda\pi + O(s^{-\frac{1}{2}}), \quad \text{as } s \to +\infty. \tag{2.27}$$

On $-\infty \leqslant s \leqslant s_0$, $L(s)$ is real, and has the value

$$L(s) = \exp\left\{-\frac{(s_0-s)^{\frac{1}{2}}}{\pi} \int_{s_0}^\infty \mathrm{d}s' \frac{\ln \eta(s')}{(s'-s_0)^{\frac{1}{2}}(s'-s)}\right\}. \tag{2.28}$$

Since $\ln \eta(s) \leqslant 0$, $L(s) \geqslant 1$ on $-\infty \leqslant s \leqslant s_0$.

If $\eta(s)$ behaves as in eqn (2.26), then for large negative s,

$$L(s) = \exp\left\{-\frac{(-s)^{\frac{1}{2}}}{\pi} \int_0^\infty \mathrm{d}s' \frac{\ln A - \lambda \ln s'}{(s')^{\frac{1}{2}}(s'-s)} + O(|s|^{-\frac{1}{2}})\right\}.$$

By eqns (I.1) and (I.2) of Appendix I this gives

$$\begin{aligned} L(s) &= \exp[-\ln A + \lambda \ln(-s) + O(|s|^{-\frac{1}{2}})] \\ &= \frac{(-s)^\lambda}{A}\{1+O((-s)^{-\frac{1}{2}})\}, \quad \text{as } s \to -\infty. \end{aligned} \tag{2.29}$$

It is now easy to see that when eqn (2.26) is satisfied,

$$L(s) \to \frac{(-s)^\lambda}{A}, \quad \text{as } |s| \to \infty, \tag{2.30}$$

where $(-s)^\lambda$ is defined (for non-integral λ) in the plane cut along $0 \leqslant s \leqslant \infty$, so that it is real on $-\infty \leqslant s \leqslant 0$.

Another simple example of $L(s)$ is the case that

$$\eta(s) \to \eta_\infty + Bs^{-(\frac{1}{2}+\epsilon)}, \quad \text{as } s \to +\infty,$$

where η_∞ and B are constants and $\eta_\infty > 0$. For large positive s,

$$\phi(s) = \frac{s^{\frac{1}{2}}}{\pi} P \int_0^\infty ds' \frac{\ln \eta_\infty}{(s')^{\frac{1}{2}}(s'-s)} + O(s^{-\frac{1}{2}}),$$

and by eqn (I.4) of Appendix I,

$$\phi(s) = O(s^{-\frac{1}{2}}), \quad \text{as } s \to +\infty.$$

For large negative s,

$$L(s) = \exp\left\{-\frac{(-s)^{\frac{1}{2}}}{\pi} \int_0^\infty ds' \frac{\ln \eta_\infty}{(s')^{\frac{1}{2}}(s'-s)} + O(s^{-\frac{1}{2}})\right\},$$

and by eqn (I.1) of Appendix I

$$L(s) = \exp\{-\ln \eta_\infty + O(s^{-\frac{1}{2}})\} = \frac{1}{\eta_\infty}\{1+O(s^{-\frac{1}{2}})\}, \quad \text{as } s \to -\infty.$$

The finite difference equation

Suppose that $\rho(s)$ is given on $-\infty \leqslant s \leqslant s_1$ (eqn (2.1)) and is kept fixed. Since $L(s)$ is real on $-\infty \leqslant s \leqslant s_1$, the equivalent elastic amplitude $\tilde{f}(s)$ of eqn (2.23) has a discontinuity across $-\infty \leqslant s \leqslant s_1$ which is given by

$$\tilde{\rho}(s) = L(s)\{\rho(s)+\text{im}(2iq(s+))^{-1}\}-\text{im}(2iq(s+))^{-1}.$$

This does not alter from one solution to another.

We assume that the equivalent elastic phases $\tilde{\alpha}(s) = \alpha(s)+\frac{1}{2}\phi(s)$ obey conditions (2.10) and (2.11), that

$$|\tilde{f}(s)| < K_3|s|^\gamma, \quad \text{as } |s| \to \infty, \tag{2.31}$$

where K_3 and γ are positive constants, and that $L(s)$ is polynomially bounded.

The equivalent elastic amplitudes are solutions to N/D equations (discussed in § 2.6). It is clear from these equations that the equivalent elastic phases $\tilde{\alpha}(s)$ are smoothly varying functions which may obey the conditions (2.10) and (2.11) even though $\alpha(s)$ and $\phi(s)$ do not separately obey these conditions. The cusps in $\phi(s)$ at channel thresholds are compensated for by similar cusps in $\alpha(s)$. For example, if $\eta(s) = 1-C(s-s_t)^{\frac{1}{2}}$ ($C > 0$) just above a threshold s_t, and $\eta(s) = 1$ just below the threshold, then $\alpha(s) = \alpha(s_t)-\frac{1}{2}C(s_t-s)^{\frac{1}{2}}$ and $\phi(s) = \phi(s_t)+C(s_t-s)^{\frac{1}{2}}$ just below s_t, while $\alpha(s) = \alpha(s_t)+O(s-s_t)$ and $\phi(s) = \phi(s_t)+O(s-s_t)$ just above s_t.

Let $f_1(s)$ and $f_2(s)$ be any two solutions of eqns (2.1) and (2.2) whose phases are $\alpha_1(s)$ and $\alpha_2(s)$. Consider the function

$$\Delta\tilde{f}(s) = \tilde{f}_2(s)-\tilde{f}_1(s).$$

It is regular across $-\infty \leqslant s \leqslant s_1$, and its only possible singularity is at $s = \infty$ and the cut $s_0 \leqslant s \leqslant \infty$. On $s_0 \leqslant s \leqslant \infty$,

$$\Delta \tilde{f}(s\pm) = \frac{\sin(\alpha_2(s)-\alpha_1(s))}{q}\exp\{\pm i(\tilde{\alpha}_1(s)+\tilde{\alpha}_2(s))\}. \tag{2.32}$$

By eqn (2.32) we see that, for $l \geqslant 1$,

$$\Delta \tilde{f}(s) \sim (s-s_0)^l, \quad \text{near } s_0 \tag{2.33}$$

(even though $\tilde{f}_1(s)$ and $\tilde{f}_2(s)$ individually need not have this property).

The function

$$\tilde{A}_{12}(s) = \exp\left\{\frac{s}{\pi}\int\limits_{s_0}^{\infty} \frac{\tilde{\alpha}_1(s')+\tilde{\alpha}_2(s')}{s'(s'-s)}\,ds'\right\} \tag{2.34}$$

has the cut $s_0 \leqslant s \leqslant \infty$, where

$$\tilde{A}_{12}(s\pm) = |\tilde{A}_{12}(s)|\exp\{\pm i(\tilde{\alpha}_1(s)+\tilde{\alpha}_2(s))\} \tag{2.35}$$

with

$$|\tilde{A}_{12}(s)| = \exp\left\{\frac{s}{\pi}P\int\limits_{s_0}^{\infty} \frac{\tilde{\alpha}_1(s')+\tilde{\alpha}_2(s')}{s'(s'-s)}\,ds'\right\}, \quad \text{for } s_0 \leqslant s \leqslant \infty. \tag{2.36}$$

The only other singularity $\tilde{A}_{12}(s)$ can have is at $s = \infty$.

It follows that
$$\Delta \tilde{f}(s) = E(s)(s-s_0)^l \tilde{A}_{12}(s), \tag{2.37}$$
where $E(s)$ is an entire function. The conditions (2.10), (2.11), and (2.31) ensure that $E(s)$ is a polynomial.

Examine what happens as $s \to +\infty$. By Theorem A in Appendix I, we have

$$|\tilde{A}_{12}(s)| \to Cs^{-\{\tilde{\alpha}_1(\infty)+\tilde{\alpha}_2(\infty)\}/\pi}, \quad \text{as } s \to +\infty \tag{2.38}$$

where C is a constant. Let the degree of the polynomial $E(s)$ be $(n-1)$ $(n \geqslant 1)$. Then by eqns (2.32), (2.37), and (2.38) we have

$$\frac{\tilde{\alpha}_1(\infty)+\tilde{\alpha}_2(\infty)}{\pi} = n+l-\tfrac{1}{2}, \tag{2.39}$$

provided
$$\tilde{\alpha}_2(\infty)-\tilde{\alpha}_1(\infty) \neq 0 \pmod{\pi}.$$

However, if $\Delta f(s) = f_2(s)-f_1(s)$ is not identically zero, but

$$\tilde{\alpha}_2(\infty)-\tilde{\alpha}_1(\infty) = 0 \pmod{\pi},$$

then remembering that $\tilde{\alpha}_1(s)$ and $\tilde{\alpha}_2(s)$ obey conditions of the form (2.11) it follows that

$$\frac{\tilde{\alpha}_1(\infty)+\tilde{\alpha}_2(\infty)}{\pi} > n+l-\tfrac{1}{2}. \tag{2.40}$$

These results give the following lemmas.

LEMMA 1. *If $f_1(s)$ and $f_2(s)$ are two distinct solutions of the p.w.d.r. then either*

$$\tilde{\alpha}_1(\infty)-\tilde{\alpha}_2(\infty) = 0 \pmod{\pi},$$

or

$$\tilde{\alpha}_1(\infty)+\tilde{\alpha}_2(\infty) = \tfrac{1}{2}\pi \pmod{\pi}.$$

LEMMA 2. *If $f_1(s)$ and $f_2(s)$ are two distinct solutions of the p.w.d.r. then*

$$\frac{\tilde{\alpha}_1(\infty)+\tilde{\alpha}_2(\infty)}{\pi} \geqslant l+\tfrac{1}{2}, \quad \text{if } \tilde{\alpha}_1(\infty) \neq \tilde{\alpha}_2(\infty) \pmod{\pi},$$

but

$$\frac{\tilde{\alpha}_1(\infty)+\tilde{\alpha}_2(\infty)}{\pi} > l+\tfrac{1}{2}, \quad \text{if } \tilde{\alpha}_1(\infty) = \tilde{\alpha}_2(\infty) \pmod{\pi}.$$

Spectra of $\tilde{\alpha}(\infty)$

Suppose we have one solution with $\tilde{\alpha}(\infty) = a$. Lemma 1 shows that *any other* solution must have $\tilde{\alpha}(\infty)$ of form

either $$\tilde{\alpha}(\infty) = a \pmod{\pi},$$

or $$\tilde{\alpha}(\infty) = -a+\tfrac{1}{2}\pi \pmod{\pi}.$$

Now we find the ($\pm$ve) integer n_0 such that

$$0 \leqslant \tfrac{1}{2}\pi-2a+n_0\pi < \pi.$$

Then $b = \frac{1}{2}\pi-a+n_0\pi$ lies in the range $a \leqslant b < a+\pi$. The values of $\tilde{\alpha}(\infty)$ which can occur must be contained in the set

$$a+n\pi, \quad b+n\pi \qquad (n = 0, \pm 1, \pm 2,\ldots).$$

This set is shown in Fig. 2.3. It is a discrete set.

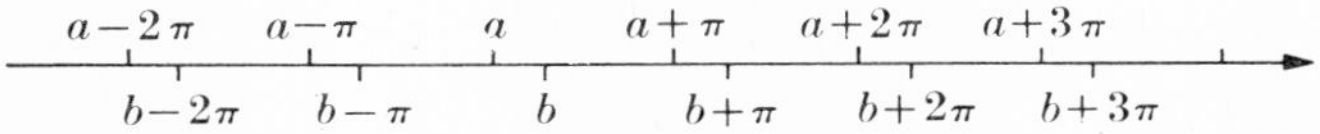

$a-2\pi$ $a-\pi$ a $a+\pi$ $a+2\pi$ $a+3\pi$

$b-2\pi$ $b-\pi$ b $b+\pi$ $b+2\pi$ $b+3\pi$

FIG. 2.3. The possible values of $\tilde{\alpha}(\infty)$ in the case that $b \neq a \pmod{\pi}$.

The main theorems

We consider only the solutions $f(s)$ of a particular p.w.d.r. for which the equivalent elastic phase shifts $\tilde{\alpha}(s)$ obey the conditions (2.10) and (2.11) of § 2.1. The solutions $f(s)$ are characterized by the functions $\tilde{\alpha}(s)$, so we can look on the set of solutions as lying in the set $\mathscr{A}$ of real functions on the interval $[s_0, \infty]$, obeying the conditions (2.10) and (2.11), We wish to know what is the largest connected region of solutions which contains a given solution $f(s)$. For the metric in $\mathscr{A}$ we use

$$d(\alpha(s), \alpha'(s)) = \sup_{s\in[s_0,\infty]} |\alpha(s)-\alpha'(s)|.$$

From the spectra of possible values of $\tilde{\alpha}(\infty)$ we can see that all solutions in the same connected region must have the same $\tilde{\alpha}(\infty)$.

If we write $$\Delta\alpha(s) = \alpha_2(s)-\alpha_1(s),$$

eqns (2.32) and (2.37) give

$$\sin(\Delta\alpha(s)) = qE(s)(s-s_0)^l|\tilde{\Lambda}_1(s)|\exp\left\{\frac{s}{\pi}P\int_{s_0}^{\infty}\frac{\Delta\alpha(s')}{s'(s'-s)}\,ds'\right\}, \tag{2.41}$$

where $\tilde{\Lambda}_1(s)$ is obtained by replacing $\alpha(s')$ in eqn (2.5) with $\tilde{\alpha}_1(s')$.

Let $f_1(s)$ be a given solution and $f_2(s)$ any solution in the same connected region. By using eqn (2.41) and $\Delta\alpha(\infty) = 0$ (plus the boundary conditions of eqns (2.10) and (2.11)), we see that for $\Delta\alpha(s) \not\equiv 0$, it is necessary that

$$n < \frac{2\tilde{\alpha}_1(\infty)}{\pi} - l + \tfrac{1}{2}, \tag{2.42}$$

where $(n-1)$ is the degree of the polynomial $E(s)$ $(n \geqslant 1)$. This provides us with a further lemma.

LEMMA 3. *If a solution $f(s)$ obeys*

$$\frac{2\tilde{\alpha}(\infty)}{\pi} \leqslant l + \tfrac{1}{2}$$

it is an isolated solution.

Now return to eqn (2.41). Suppose that the solution $f_1(s)$ is given. Let p be the integer determined by

$$p < \frac{2\tilde{\alpha}_1(\infty)}{\pi} - l + \tfrac{1}{2} \leqslant p+1.$$

Lemma 3 shows that if $p < 1$ then $f_1(s)$ is an isolated solution. If $p \geqslant 1$ it can be shown that, given any real polynomial $\xi(s)$ of degree $(p-1)$, there exists a real number $\lambda_1 > 0$, such that the equation

$$\sin(\Delta\alpha(s,\lambda)) = \lambda q \xi(s)(s-s_0)^l |\tilde{A}_1(s)| \exp\left\{\frac{s}{\pi} P \int_{s_0}^{\infty} \frac{\Delta\alpha(s',\lambda)}{s'(s'-s)}\, ds'\right\} \tag{2.43}$$

has a solution $\Delta\alpha(s,\lambda)$ for $|\lambda| < \lambda_1$. The different solutions obtained by varying λ continuously in $-\lambda_1 < \lambda < \lambda_1$ are connected in $\mathscr{A}$.

In Appendix II we show that the solutions of eqn (2.43) are given by

$$\tan(\Delta\alpha(s,\lambda)) = \lambda h(s)\left\{1 - \frac{\lambda s}{\pi} P \int_{s_0}^{\infty} \frac{h(s')\, ds'}{s'(s'-s)}\right\}^{-1}, \tag{2.44}$$

where

$$h(s) = q\xi(s)(s-s_0)^l |\tilde{A}_1(s)|. \tag{2.45}$$

Furthermore, λ_1 is the largest value of $|\lambda|$ for which the function

$$G(s,\lambda) = 1 - \frac{\lambda s}{\pi} \int_{s_0}^{\infty} \frac{h(s')\, ds'}{s'(s'-s)}$$

does not vanish for all $|\lambda| < \lambda_1$ for any value of the complex variable s. These results give us Theorem 1.

THEOREM 1. *Let $f(s)$ be any solution of the given p.w.d.r. and let p be the integer defined by*

$$p < \frac{2\tilde{\alpha}(\infty)}{\pi} - l + \tfrac{1}{2} \leqslant p+1.$$

Then if $p < 1$, $f(s)$ is an isolated solution, whereas if $p \geqslant 1$, $f(s)$ is contained in a p-dimensional manifold of solutions. The other solutions of the manifold have phase shifts $\alpha(s)+\Delta\alpha(s,\lambda)$, where $\Delta\alpha(s,\lambda)$ is given by eqns (2.44) *and* (2.45).

Let $f_1(s)$ be an isolated solution and consider first the case that

$$\frac{2\tilde{\alpha}_1(\infty)}{\pi} < l + \tfrac{1}{2}.$$

Let $f_2(s)$ be any other solution of the given p.w.d.r. Then

$$l + \tfrac{1}{2} + \frac{2\tilde{\alpha}_2(\infty)}{\pi} > 2\,\frac{\tilde{\alpha}_1(\infty)+\tilde{\alpha}_2(\infty)}{\pi} \geqslant 2(l+\tfrac{1}{2}),$$

where Lemma 2 is used in the last step. So

$$\frac{2\tilde{\alpha}_2(\infty)}{\pi} > l + \tfrac{1}{2}. \tag{2.46}$$

Equation (2.46) and Theorem 1 show that $f_2(s)$ is not an isolated solution. Suppose that $f_1(s)$ is an isolated solution with

$$\frac{2\tilde{\alpha}_1(\infty)}{\pi} = l + \tfrac{1}{2}.$$

Lemma 2 requires that
$$\frac{2\tilde{\alpha}_2(\infty)}{\pi} \geqslant l + \tfrac{1}{2}.$$

and the last part of Lemma 2 now excludes $\tilde{\alpha}_1(\infty) = \tilde{\alpha}_2(\infty)$, so once again eqn (2.46) holds. This gives us a further theorem.

THEOREM 2. *There can be at most one isolated solution of a given p.w.d.r.*

Left-hand cuts can be constructed such that there is no isolated solution.

Finally we again emphasize that we have assumed that the equivalent elastic phases $\tilde{\alpha}(s)$ obey conditions (2.10) and (2.11) and that condition (2.31) is obeyed.

2.4. Several bound states and Levinson's theorem

In § 2.2 we discussed uniqueness in the case of one bound state, using the simple method. We shall now generalize to a finite number of bound states, and at the same time we shall apply the improved method of § 2.3.

Suppose that the p.w.a. $f(s)$ has bound state poles at $B_1, B_2,..., B_n$, having the residues $\Gamma_1, \Gamma_2,..., \Gamma_n$ respectively. Let $f(s)$ have the phase $\alpha(s)$ on the physical cut $s_0 \leqslant s \leqslant \infty$. The equivalent elastic amplitude $\tilde{f}(s)$ is defined as in eqn (2.23). It has the phase $\tilde{\alpha}(s)$ defined as in eqn (2.25), and bound state poles at B_i $(i = 1,..., n)$. Remembering that $L(s) \geqslant 1$ on $-\infty < s < s_0$, we find their residues to be

$$\tilde{\Gamma}_i = L(B_i)\Gamma_i \quad (i = 1, 2,..., n).$$

It is assumed that $\tilde{\alpha}(s)$ obeys conditions (2.10) and (2.11) and that $\tilde{f}(s)$ obeys the condition (2.31).

First we apply the finite difference method to any two solutions $f_1(s)$, $f_2(s)$. Let $f_1(s)$ have the phase $\alpha_1(s)$ and have n_1 bound states whose poles are at B_i and have residues Γ_i $(i = 1,..., n_1)$. Let $f_2(s)$ have the phase $\alpha_2(s)$ and have n_2 bound states whose poles are at B'_i and have residues Γ'_i $(i = 1,..., n_2)$.

Consider the function

$$\Delta\tilde{f}(s) = \tilde{f}_2(s) - \tilde{f}_1(s).$$

On the physical cut $\Delta\tilde{f}(s+)$ has the phase factor $\exp\{i(\tilde{\alpha}_1(s)+\tilde{\alpha}_2(s))\}$. The same is true of the function

$$\Delta\tilde{f}(s) \prod_{i=1}^{n_1} (s-B_i) \prod_{i=1}^{n_2} (s-B'_i).$$

This function has no singularities other than the cut $s_0 \leqslant s \leqslant \infty$, and it is bounded by a power of $|s|$ as $|s| \to \infty$. Therefore we have the relation

$$\Delta\tilde{f}(s) \prod_{i=1}^{n_1} (s-B_i) \prod_{i=1}^{n_2} (s-B'_i) = (s-s_0)^l \xi(s) \tilde{\Lambda}_{12}(s), \tag{2.47}$$

where $\tilde{\Lambda}_{12}(s)$ is given by eqn (2.34) and $\xi(s)$ is a polynomial. Also l is the orbital angular momentum and the factor $(s-s_0)^l$ appears on the right of eqn (2.47) because $\Delta\tilde{f}(s) \sim (s-s_0)^l$ near s_0.

By arguments similar to those in § 2.3 we now have a further lemma.

LEMMA 4. *If $f_1(s)$ and $f_2(s)$ are two distinct solutions of the p.w.d.r. with n_1 and n_2 bound states, then*

either $$\tilde{\alpha}_1(\infty) - \tilde{\alpha}_2(\infty) = 0 \pmod{\pi},$$

or $$\tilde{\alpha}_1(\infty) + \tilde{\alpha}_2(\infty) = \tfrac{1}{2}\pi \pmod{\pi}.$$

Further,

$$\frac{\tilde{\alpha}_1(\infty)}{\pi} + n_1 + \frac{\tilde{\alpha}_2(\infty)}{\pi} + n_2 \geqslant l + \tfrac{1}{2}, \quad \textit{if } \tilde{\alpha}_1(\infty) \neq \tilde{\alpha}_2(\infty) \pmod{\pi},$$

and $$\frac{\tilde{\alpha}_1(\infty)}{\pi} + n_1 + \frac{\tilde{\alpha}_2(\infty)}{\pi} + n_2 > l + \tfrac{1}{2}, \quad \textit{if } \tilde{\alpha}_1(\infty) = \tilde{\alpha}_2(\infty) \pmod{\pi}.$$

We consider the set of solutions $f(s)$ lying in the metric space $\mathscr{A}$ defined as in § 2.3. In Appendix II we examine the structure of the set of solutions in the neighbourhood of a given solution in $\mathscr{A}$. We then find Theorem 3.

THEOREM 3. *Let $f(s)$ be a solution of the given p.w.d.r. with n bound states at the positions B_i ($i = 1, 2,..., n$). Let $\tilde{\alpha}(s)$ be the equivalent elastic phase of $f(s)$. The index of $f(s)$ is the integer p given by*

$$p < 2\left(\frac{\tilde{\alpha}(\infty)}{\pi}+n\right)-l+\tfrac{1}{2} \leqslant p+1. \tag{2.48}$$

(*a*) *There is at most one solution with $p \leqslant 0$ (the isolated solution).*

(*b*) *If $p > 0$, then $f(s)$ is contained in a p-dimensional manifold of solutions having the same index p.*

(*c*) *There is a neighbourhood of $f(s)$ which contains no solutions of index less than p.*

(*d*) *Let m be an integer such that $m \geqslant n$ and $p+m-n > 0$. In each neighbourhood of $f(s)$ there is a $(p+2(m-n))$-dimensional manifold of solutions with m bound states and index $p+2(m-n)$.*

From (*d*) it follows that arbitrarily close to any phase shift there are other phase shift solutions. However, for the isolated solution these other solutions have more bound states.

An interesting consequence of the last part of Lemma 4 and Theorem 3 is that if for the isolated solution

$$\frac{\tilde{\alpha}(\infty)}{\pi}+n = \tfrac{1}{2}(l+\tfrac{1}{2})-\tfrac{1}{2}\Delta, \quad \text{where } \Delta > 0,$$

then the interval in the value $(\tilde{\alpha}(\infty)/\pi+n)$ before the next solution occurs is at least Δ.

Levinson's theorem

Levinson's theorem applies to non-relativistic Schrödinger theory for a (real) central potential $V(r)$. The theorem has been proved ([9], [10], [11]) for potentials obeying

$$\left.\begin{aligned} \int_0^\infty r|V(r)|\,dr &< \infty \\ \int_0^\infty r^2|V(r)|\,dr &< \infty \end{aligned}\right\}. \tag{2.49}$$

The method is restricted to elastic scattering. The method of proof has been generalized by Jauch [12], who shows that the theorem can be deduced if it is known that the set of eigenfunctions of the total Hamiltonian for the elastic two-body problem is complete.

We shall state the theorem in terms of Jost's function [11]. If k is the asymptotic momentum of the incoming spherical wave, the S-matrix element for angular momentum l may be written

$$S_l(k) = g_l(k)/g_l(-k),$$

where $g_l(k)$ is Jost's function (see Hamilton [13], p. 34, for examples). We define the phase by

$$\delta_l(k) = \arg g_l(k), \quad \text{for } 0 \leqslant k \leqslant \infty.$$

It can be shown that under the conditions (2.49) on the potential, $\delta_l(k) \to 0$ as $k \to \infty$. With this definition we may have $\delta_l(0) \neq 0$.

LEVINSON'S THEOREM.

(*a*) *If* $g_l(0) \neq 0$, *then*

$$\delta_l(0) - \delta_l(\infty) = n_B \pi \tag{2.50}$$

where n_B *is the number of bound states.*

(*b*) *If* $g_l(0) = 0$ *and* $l > 0$, *there is a discrete bound state at* $k = 0$ *and eqn* (2.50) *holds when this state is included in the number* n_B.

(*c*) *If* $g_0(0) = 0$ (*i.e. an S-wave*) *the wave function is not normalized for* $k = 0$, *and* $k = 0$ *does not correspond to a discrete bound state. In this case eqn* (2.50) *is to be replaced by*

$$\delta_0(0) - \delta_0(\infty) = (n_B + \tfrac{1}{2})\pi, \tag{2.51}$$

n_B *being the number of discrete bound states.*

For comparison with our results, remember that we took $\alpha(s_0) = 0$, so $\alpha(\infty)$ is to be compared with $\delta_l(\infty) - \delta_l(0)$. In our notation (ignoring the singular S-wave case giving eqn (2.51), as it is not relevant to our discussion) eqn (2.50) is

$$\frac{\alpha(\infty)}{\pi} + n_B = 0. \tag{2.52}$$

Theorem 3 applied to elastic scattering shows that if

$$\frac{\alpha(\infty)}{\pi} + n_B \leqslant \tfrac{1}{2}(l + \tfrac{1}{2}), \tag{2.53}$$

the solution is isolated. Thus the partial wave solutions obtained from potentials obeying the conditions (2.49) are isolated solutions.

It has been shown by Atkinson *et al.* [14], [15] that CDD poles appear in some cases where multichannel problems are treated by the single channel method [3]. However, it is reasonable to believe that in a large number of elementary particle problems the isolated solution is the relevant solution. For this reason in Chapters 4–6 we shall examine the location of the second sheet poles of the isolated solutions for various attractive P-waves.

2.5. Behaviour at infinity and subtraction

In studying the arbitrariness of solutions of p.w.d.r. we have (in § 2.3) used condition (2.31), stating

$$|f(s)| < K_3|s|^{\gamma}, \quad \text{as } |s| \to \infty, \tag{2.31}$$

where K_3 and γ are any positive constants.

It is possible to replace this condition by two conditions which may be more convenient. We use the theorem of Phragmén and Lindelöf in the form discussed on pp. 177–8 of Titchmarsh [16].

Let $f(z)$ be an analytic function of $z = r\exp(i\theta)$, regular in the region D between two straight lines making an angle π/α at the origin and continuous on the lines themselves. Suppose that

$$|f(z)| \leqslant M$$

on the lines, and that as $r \to \infty$

$$f(z) = O(\exp(r^{\beta})),$$

where as usual the symbol $O(x)$ has its strict meaning of order of x or smaller, and where $\beta < \alpha$, uniformly in the angle. Then actually the inequality

$$|f(z)| \leqslant M$$

holds throughout D.

We have
$$|f(s)| = O(s^{-\frac{1}{2}}), \quad \text{as } s \to +\infty.$$

If we assume that

$$|f(s)| < K|s|^{\gamma}, \quad \text{as } s \to -\infty,$$

and assume that for some $\delta > 0$,

$$|f(s)| = O(\exp(|s|^{1-\delta})), \quad \text{as } |s| \to \infty, \tag{2.54}$$

then condition (2.31) follows.

The number of subtractions

Kinoshita [17] has considered the number of subtractions required in an elastic p.w.a. $A_l(s)$ (defined as in Chapter 1) where $|A_l(s)|$ is bounded by a constant as $s \to +\infty$ (see Chapter 5). He considers various fairly general conditions. His results are given in Theorem 4.

THEOREM 4. *A p.w. projection $A_l(s)$ of orbital angular momentum l ($l \geqslant 0$) satisfies a dispersion relation with at most one subtraction if*

(*a*) $$|A_l(s)| < \exp(C|s|^{1-\delta}), \quad \textit{as } |s| \to \infty,$$

where C and δ are positive constants† *and*

(*b*) *the number of changes of sign of* $\operatorname{im} A_l(s+)$ *on the left-hand cut* $-\infty \leqslant s \leqslant s_1$ *is finite.*

Another pair of sufficient conditions is

(i) $$|A_l(s)| < \exp(C(\ln|s|)^{2-\delta}), \quad as\ |s| \to \infty,$$

and

(ii) *the number of times* $\operatorname{im} A_l(s+)$ *changes its sign on the interval* (s, s_1) *does not exceed*

$$C'(\ln|s|)^{1-\delta}, \quad as\ s \to -\infty.$$

Here C, C' and δ are positive constants.

Notice that if $l \geqslant 1$ the subtraction can be made at the physical threshold and no arbitrary constant is introduced. Since $q^2 \sim s$ for $|s| \to \infty$, no subtraction is needed when we use the reduced p.w.a. $A_l(s)/q^{2l}$ for $l \geqslant 1$.

It is easy to adapt Kinoshita's method to the general elastic p.w.a. $f_l(s)$ (specified as in eqn (1.1) with $\eta \equiv 1$). As a result we obtain Theorem 5.

THEOREM 5. *A p.w.a.* $f_l(s)$, *where* $l \geqslant 0$, *obeys an unsubtracted dispersion relation if*

(1) $$|f_l(s)| < \exp(C|s|^{1-\delta}), \quad as\ |s| \to \infty,$$

where C and δ are positive constants†, *and*

(2) *the number of times* $\operatorname{im} f_l(s+)$ *changes its sign on the left-hand cut* $-\infty \leqslant s \leqslant s_1$ *is finite.*

Thus under the fairly general conditions (1) and (2) the unsubtracted p.w.d.r. of eqn (2.1) is valid as it stands, except that if bound states exist we must add their poles to the right of the equation. From now on we shall assume that the equivalent elastic amplitude $\tilde{f}_l(s)$ of eqn (2.23) obeys the conditions (1) and (2) above.

The value of $\tilde{\alpha}(\infty)$

We use eqn (2.1) for the isolated equivalent elastic amplitude $\tilde{f}_l(s)$. If the phase shift $\tilde{\alpha}(s)$ obeys the conditions (2.10) and (2.11), then by using a method like that in Limit theorem A (Appendix I) we can prove that

$$\lim_{s \to +\infty} \frac{q}{\pi} P \int_{s_0}^{\infty} \frac{\operatorname{im} \tilde{f}_l(s'+)\, ds'}{s'-s} = 0.$$

Hence
$$\lim_{s \to +\infty} 2q \operatorname{re} \tilde{f}_l(s+) = \sin(2\tilde{\alpha}(\infty))$$
$$= \lim_{s \to +\infty} \frac{2q}{\pi} \int_{-\infty}^{s_1} \frac{\operatorname{im} \tilde{f}_l(s'+)}{s'-s}\, ds'. \tag{2.55}$$

† One may have to assume also that the number of zeros and poles is finite.

Therefore $$\int_{-\infty}^{s_1} \frac{\operatorname{im} \tilde{f}_l(s'+)}{s'-s}\, ds' \to \pi \frac{\sin(2\tilde{\alpha}(\infty))}{s^{\frac{1}{2}}}, \quad \text{as } s \to +\infty. \tag{2.56}$$

For a *truncated left-hand cut* (for which $\operatorname{im} \tilde{f}_l(s+) = 0$ for $-\infty \leqslant s \leqslant s_2$, where $s_2 < s_1$) the left-hand side of eqn (2.56) is $O(s^{-1})$ as $s \to +\infty$, therefore

$$\sin(2\tilde{\alpha}(\infty)) = 0,$$

i.e.

$$\tilde{\alpha}(\infty) = 0 \quad (\text{mod } \tfrac{1}{2}\pi).$$

If $\tilde{\alpha}(\infty)$ is not zero or a multiple of $\frac{1}{2}\pi$, then the left-hand cut is not truncated. In this case we can apply a theorem of Hardy and Littlewood [18].

THEOREM 6. *Suppose $g(t)$ is positive and integrable over every range $(0, T)$ and*

$$\frac{g(t)}{t+x}$$

is integrable over $(0, \infty)$ for some (and so all) positive x. Suppose further that

$$h(x) = \int_0^\infty \frac{g(t)\, dt}{t+x} \to H/x^\sigma, \quad \textit{as } x \to +\infty, \tag{2.57}$$

where $0 < \sigma < 1$ and $H > 0$. Then

$$G(t) = \int_0^t g(u)\, du \to \frac{H \sin(\pi\sigma)}{(1-\sigma)\pi} t^{(1-\sigma)}, \quad \textit{as } t \to +\infty. \tag{2.58}$$

The result is still true for $H = 0$ if we interpret eqn (2.57) to mean $x^\sigma h(x) \to 0$ as $x \to +\infty$, and eqn (2.58) to mean $t^{\sigma-1}G(t) \to 0$ as $t \to +\infty$.

The same results hold if $g(t)$ has at most a finite number of changes of sign for positive t. Applying this to eqn (2.56) we have Theorem 7.

THEOREM 7. *If $\tilde{f}_l(s)$ obeys conditions (1) and (2) of Theorem 5 and if $\tilde{\alpha}(s)$ obeys conditions (2.10) and (2.11), and if $\tilde{\alpha}(\infty) = 0$ (mod $\frac{1}{2}\pi$), then*

$$\frac{1}{\sqrt{(-s)}} \int_s^{s_1} \operatorname{im} \tilde{f}_l(s'+)\, ds' \to 0, \quad \textit{as } s \to -\infty; \tag{2.59}$$

while if $\tilde{\alpha}(\infty) \neq 0$ (mod $\frac{1}{2}\pi$), then

$$\int_s^{s_1} \operatorname{im} \tilde{f}_l(s'+)\, ds' \to -2\sin(2\tilde{\alpha}(\infty))\sqrt{(-s)}, \quad \textit{as } s \to -\infty. \tag{2.60}$$

Applications: power law behaviour

Suppose that the conditions of Theorem 7 are satisfied, and suppose in addition that

$$\operatorname{im} \tilde{f}_l(s+) \sim (-s)^{\gamma}, \quad \text{as } s \to -\infty,$$

where γ is a constant. From Theorem 7 it follows that only $\gamma \leqslant -\frac{1}{2}$ is possible. If $\gamma < -\frac{1}{2}$ then $\tilde{\alpha}(\infty) = 0 \pmod{\frac{1}{2}\pi}$, and only $\gamma = -\frac{1}{2}$ can give $\tilde{\alpha}(\infty) \neq 0 \pmod{\frac{1}{2}\pi}$.

If $\operatorname{im} \tilde{f}_l(s+)$ is sufficiently smoothly behaved as $s \to -\infty$ the converse is true, namely $\tilde{\alpha}(\infty) \neq 0 \pmod{\frac{1}{2}\pi}$ implies $\operatorname{im} \tilde{f}_l(s+) \sim (-s)^{\frac{1}{2}}$ as $s \to -\infty$.

Truncated left-hand cut and $\tilde{\alpha}(\infty) = 0 \pmod{\pi}$

We can use another method to find how $\tilde{\alpha}(s)$ approaches $\tilde{\alpha}(\infty)$ in this case. We assume that $\tilde{\alpha}(s)$ obeys conditions (2.10) and (2.11) and define

$$\begin{aligned}\mathscr{D}(s) &= (\tilde{A}(s))^{-\frac{1}{2}} \\ &= \exp\left\{-\frac{s}{\pi}\int_{s_0}^{\infty} \frac{\tilde{\alpha}(s')\,ds'}{s'(s'-s)}\right\}\end{aligned} \tag{2.61}$$

(compare with eqn (2.5)). The only singularity of $\mathscr{D}(s)$ is the cut $s_0 \leqslant s \leqslant \infty$. On it

$$\mathscr{D}(s\pm) = |\mathscr{D}(s)| \exp(\mp i\tilde{\alpha}(s)). \tag{2.62}$$

Thus the function $\tilde{f}(s)\mathscr{D}(s)$ is regular except for the left-hand cut $s_2 \leqslant s \leqslant s_1$ and possibly bound state poles at $s = \beta_i$. So we can write

$$\tilde{f}(s)\mathscr{D}(s) = \frac{1}{\pi}\int_{s_2}^{s_1} ds' \frac{\tilde{\rho}(s')\mathscr{D}(s')}{s'-s} + \sum_{i=1}^{n_B} \frac{\tilde{\Gamma}_i \mathscr{D}(\beta_i)}{s-\beta_i} + E(s). \tag{2.63}$$

Here $E(s)$ is a real polynomial, $\tilde{\Gamma}_i$ are real constants, and $\tilde{\rho}(s) = \operatorname{im} \tilde{f}(s+)$, for $s_2 \leqslant s \leqslant s_1$. Equation (2.63) shows that for large $|s|$, $\tilde{f}\mathscr{D}$ behaves like s^m where m is a positive or negative integer or zero. On $s_0 \leqslant s \leqslant \infty$, $\tilde{f}\mathscr{D}$ is regular, and

$$\tilde{f}(s)\mathscr{D}(s) = \frac{\sin\tilde{\alpha}(s)}{q}\,|\mathscr{D}(s)|, \quad \text{for } s_0 \leqslant s \leqslant \infty. \tag{2.64}$$

By eqns (2.61), (2.64), and (2.12) we get

$$\frac{\sin\tilde{\alpha}(s)}{q} \sim s^{\{m-\tilde{\alpha}(\infty)/\pi\}}, \quad \text{as } s \to +\infty.$$

Therefore, if $\tilde{\alpha}(\infty) = 0 \pmod{\pi}$,

$$\tilde{\alpha}(s) - \tilde{\alpha}(\infty) \to Cs^{-\beta}, \quad \text{as } s \to +\infty, \tag{2.65}$$

where C is a constant and

$$\beta = n+\tfrac{1}{2} \quad (n = 0, 1, 2,...). \tag{2.66}$$

2.6. Relation to CDD poles (case of S-waves)

There is a simple relation between the manifolds of p.w.a. solutions and CDD poles. Since we wish to be brief it is sufficient to look at S-waves. We shall allow inelasticity and we use the function $\mathscr{D}(s)$ defined by eqn (2.61). In the introductory remarks we shall consider the general left-hand cut, but we shall restrict ourselves to the truncated left-hand cut when we discuss the number of arbitrary parameters.

It follows from eqn (2.62) that the function

$$\mathscr{N}(s) = \tilde{f}(s)\mathscr{D}(s) \tag{2.67}$$

is regular on $s_0 \leqslant s \leqslant \infty$, and the only singularities of $\mathscr{N}(s)$ are the unphysical cut of $\tilde{f}(s)$, $-\infty \leqslant s \leqslant s_1$, plus any bound state poles of $\tilde{f}(s)$.

Equation (2.67) does not give the most general form of the functions $N(s)$ and $D(s)$. We can use

$$D(s) = \frac{P(s)}{Q(s)}\mathscr{D}(s), \tag{2.68}$$

where $P(s)$, $Q(s)$ are real polynomials in s, having no common factors. We define $N(s)$ by

$$N(s) = \tilde{f}(s)D(s)$$
$$= \tilde{f}(s)\frac{P(s)}{Q(s)}\mathscr{D}(s). \tag{2.69}$$

The function $N(s)$ has the cut $-\infty \leqslant s \leqslant s_1$, but it does not have the cut $s_0 \leqslant s \leqslant \infty$. We further require that $N(s)$ has no other singularities. Thus $P(s)$ must contain factors $(s-B_i)$ which vanish at the bound states B_i. Suppose there are n_B bound states. We use

$$P(s) = P_B(s) = \prod_{i=1}^{n_B}(s-B_i).$$

Nothing is gained by having further factors in $P(s)$, since in that case $N(s)$ would be subject to the unnecessary restriction that it must vanish at these extra zeros of $P(s)$.

So eqns (2.68) and (2.69) become

$$D(s) = \frac{P_B(s)}{Q(s)}\mathscr{D}(s), \tag{2.70}$$

$$N(s) = \tilde{f}(s)\frac{P_B(s)}{Q(s)}\mathscr{D}(s). \tag{2.71}$$

Let the degree of $Q(s)$ be n_c. The condition that $N(s)$ should have no singularities other than the unphysical cut $-\infty \leqslant s \leqslant s_1$ requires $\tilde{f}(s)$ to vanish at the zeros of $Q(s)$. Such zeros can lie on the physical cut $s_0 \leqslant s \leqslant \infty$ at the points (or at some of the points) where $\tilde{\alpha}(s) = 0 \pmod{\pi}$. The zeros can also lie elsewhere.

The function $D(s)$ has poles at the zeros of $Q(s)$; they are called the CDD poles after the original discoverers of the arbitrariness problem [1]. Using the unitarity equation

$$\mathrm{im}(\tilde{f}^{-1}(s+)) = -q, \quad \text{for } s_0 \leqslant s \leqslant \infty,$$

we can write down the dispersion relations for $N(s)$ and $D(s)$,

$$N(s) = \frac{1}{\pi}\int_{-\infty}^{s_1} \mathrm{d}s' \, \frac{D(s')\mathrm{im}\,\tilde{f}(s'+)}{s'-s}, \tag{2.72}$$

$$D(s) = 1+\frac{s-\bar{s}}{\pi}\left\{\sum_{j=1}^{n_c} \frac{c_j}{s-s_j} - \int_{s_0}^{\infty} \mathrm{d}s' \, \frac{q'N(s')}{(s'-\bar{s})(s'-s)}\right\} \tag{2.73}$$

where s_j ($j = 1,..., n_c$) are the CDD poles and c_j are arbitrary constants. The equation for $D(s)$ has been subtracted at $\bar{s}$ where $-\infty < \bar{s} < s_0$, but of course (assuming $\tilde{f}(s)$ obeys eqn (2.31)) we may need further subtractions in eqns (2.72) and (2.73). Since $Q(s)$ is a real polynomial and $D(s^*) = (D(s))^*$ there are *two* arbitrary constants (c_j, s_j) associated with each CDD pole.

Counting the arbitrary parameters (truncated left-hand cut)

(*a*) $\tilde{\alpha}(\infty) = \frac{1}{2}\pi \pmod{\pi}$

Consider the N and D equations like eqns (2.72) and (2.73), but with as many subtractions as we wish. On account of the truncated left-hand cut the function $N = \tilde{f}D$ is of the form s^m for large $|s|$, where m is an integer (positive, negative, or zero). Letting $s \to +\infty$, remembering that n_c is the degree of $Q(s)$, and using eqns (2.69) and (2.64),†

$$m = -\tfrac{1}{2}+\frac{\tilde{\alpha}(\infty)}{\pi}+n_B-n_c \tag{2.74}$$

or

$$2\left(\frac{\tilde{\alpha}(\infty)}{\pi}+n_B\right) = 2m+2n_c+1. \tag{2.75}$$

By eqn (2.48) the degree of arbitrariness p of the solution $\tilde{f}(s)$ is given by

$$p < 2\left(\frac{\tilde{\alpha}(\infty)}{\pi}+n_B\right)+\tfrac{1}{2} \leqslant p+1.$$

† For simplicity we consider only the case where $D(s)$ is given by eqn (2.70).

Equation (2.75) gives $$p = 2m+1+2n_c. \tag{2.76}$$

Since $\tilde{f}(s) \to i/q$ and $N(s) \sim s^m$, as $s \to +\infty$,

we require $$\operatorname{re} D(s)/s^{(m+\frac{1}{2})} \to 0, \quad \text{as } s \to +\infty. \tag{2.77}$$

If $m = -1$

Equations (2.72) and (2.73) can be used as they stand without further subtractions. By eqn (2.77) we want $s^{\frac{1}{2}} \operatorname{re} D(s) \to 0$ as $s \to +\infty$. This imposes one constraint on the parameters c_j appearing in eqn (2.73).† Thus N and D contain $(2n_c-1)$ arbitrary parameters. This agrees with the value of p in eqn (2.76).

If $m = -2$

Again eqns (2.72) and (2.73) can be used as they stand. Choose the c_j, s_j so that

$$\int_{s_2}^{s_1} D(s) \operatorname{im} \tilde{f}(s+)\, ds = 0.$$

This ensures that $N(s) \sim s^{-2}$ as $s \to \infty$, and imposes one constraint on the c_j, s_j. Now write eqn (2.73) in the form

$$\operatorname{re} D(s) = 1 + \frac{s-\bar{s}}{\pi} \sum_{j=1}^{n_c} \frac{c_j}{s-s_j} + \frac{1}{\pi} \int_{s_0}^{\infty} \frac{q' N(s')}{s'-\bar{s}}\, ds' + $$

$$+ \frac{1}{\pi s} \int_{s_0}^{\infty} q' N(s')\, ds' - \frac{1}{\pi s} P \int_{s_0}^{\infty} \frac{q' . s' . N(s')}{s'-s}\, ds'. \tag{2.78}$$

By eqn (I.4) of Appendix I the last term on the right of eqn (2.78) is $O(s^{-2})$, as $s \to +\infty$. Two further constraints on the c_j, s_j can make the other terms on the right of eqn (2.78) of order $O(s^{-2})$; then $\operatorname{re} D(s) = O(s^{-2})$, as $s \to +\infty$. Hence the number of arbitrary parameters in N and D is $(2n_c-3)$; this agrees with p in eqn (2.76).

If $m \geqslant 0$

There have to be $(m+1)$ subtractions in both the N and D equations used in place of eqns (2.72) and (2.73). One parameter is determined by $D(\bar{s}) = 1$, so there are $(2m+1)$ arbitrary subtraction parameters (the positions, such as $\bar{s}$, where the subtractions are made are fixed). Adding the $2n_c$ CDD parameters we have $2m+1+2n_c$ arbitrary parameters in N and D, in agreement with eqn (2.76).

† Alternatively an unsubtracted equation can be used for $D(s)$ in place of eqn (2.73). One constraint then ensures $D(\bar{s}) = 1$. We impose $D(\bar{s}) = 1$, since it is the ratio N/D which matters.

(*b*) $\tilde{\alpha}(\infty) = 0 \pmod{\pi}$

The behaviour of $\tilde{\alpha}(s)$ as $s \to +\infty$ is given by eqn (2.65). By eqns (2.69) and (2.64), as $s \to \infty$, $N(s) \sim s^m$ with

$$m = -\tfrac{1}{2}-\beta+\frac{\tilde{\alpha}(\infty)}{\pi}+n_B-n_c, \tag{2.79}$$

and β is given by eqn (2.66). We must be particularly careful here in deducing the degree of arbitrariness p of the solutions $\tilde{f}(s)$. This is because our general theorems on the dimensions of the manifold of solutions relate to solutions having identical $\tilde{\alpha}(\infty)$, but they do not specify the rate at which $\tilde{\alpha}(s)$ approaches $\tilde{\alpha}(\infty)$ as $s \to +\infty$. It is to be expected that when this rate is specified we may get a sub-manifold having lower dimension than p.

Clearly

$$\tilde{f}(s) \to (\text{real constant}).s^{-(n+1)}, \quad \text{as } s \to +\infty,$$

where $n = 0, 1, 2, \ldots$. Thus

$$\left.\begin{aligned} N(s) &\sim s^m \\ \operatorname{re} D(s) &\sim s^{(n+m+1)} \\ \operatorname{im} D(s) &\sim s^{(m+\frac{1}{2})} \end{aligned}\right\} \quad \text{as } s \to +\infty. \tag{2.80}$$

What we might call the 'normal' case is $\beta = \frac{1}{2}$; that is, there is the least possible restriction on the manner in which $\tilde{\alpha}(s)$ tends to $\tilde{\alpha}(\infty)$. For $\beta = \frac{1}{2}$, using eqn (2.79) and the general result in eqn (2.48), the degree of arbitrariness p is

$$p = 2(m+n_c+1).$$

For $\beta = n+\frac{1}{2}$ $(n \geqslant 1)$ we would expect that the degree of arbitrariness would be the value given by eqn (2.48), *minus* n; this is based on the assumption that the extra restrictions required to get solutions $\tilde{f}(s)$ having this value of β will reduce the dimensions of the manifold by n. Using eqns (2.79) and (2.48) we now obtain

$$\begin{aligned} p &= 2(m+n+n_c+1)-n \\ &= 2(m+n_c+1)+n. \end{aligned} \tag{2.81}$$

We can use the N and D equations to show that this is indeed correct. We shall merely give one case as an example.

Case $m = -1$

Equation (2.80) shows that the unsubtracted eqn (2.72) is satisfactory for $N(s)$. For $n \geqslant 1$, eqn (2.73) for $D(s)$ must be replaced by an equation having $(n+1)$ subtractions in all. One of these is specified by $D(\bar{s}) = 1$, so the number of arbitrary constants in $N(s)$ and $D(s)$ is $2n_c+n$.

This confirms eqn (2.81) for $m = -1$. It is easy to construct similar arguments for the cases $m \leqslant -2$ and $m \geqslant 0$, and to show that in all cases eqn (2.81) gives the degree of arbitrariness when $\tilde{\alpha}(\infty) = 0 \pmod{\pi}$ and $\beta = n+\frac{1}{2}$.

Comments

(a) CDD poles or subtractions

Any degree of arbitrariness p ($p > 0$) can be maintained even if the number n_c of CDD poles is reduced, provided that the number of subtractions is correspondingly increased [19].

(b) Truncated left-hand cut and potentials

The isolated elastic S-wave solution $f_0(s)$ for a truncated left-hand cut is obtained from eqns (2.72) and (2.73) on putting all $c_i = 0$. If the left-hand cut discontinuity $\operatorname{im} f(s+)$ is such that the solution of these equations does *not* obey the sum rule

$$1+\frac{1}{\pi}\int_{s_0}^{\infty} \mathrm{d}s' \frac{q'N(s')}{s'-\bar{s}} = 0 \tag{2.82}$$

(and in the normal case it does not), then by eqn (2.73)

$$|D(s)| = O(1), \quad \text{as } |s| \to \infty.$$

Hence, by eqn (2.70),

$$\frac{\alpha(\infty)}{\pi}+n_B = 0.$$

An isolated elastic P-wave $f_1(s)$ with a truncated left-hand cut is the solution of the N and D equations:

$$F(s) = f_1(s)/q^2 = N/D;$$

$$N(s) = \frac{1}{\pi}\int_{s_2}^{s_1} \mathrm{d}s' \frac{D(s')\operatorname{im} F(s'+)}{s'-s}; \tag{2.83}$$

$$D(s) = 1-\frac{s-s_0}{\pi}\int_{s_0}^{\infty} \mathrm{d}s' \frac{q'^3N(s')}{(s'-s_0)(s'-s)}. \tag{2.84}$$

Now $\qquad \operatorname{im} D(s+) = -q^3N(s) \quad \text{for } s_0 \leqslant s \leqslant \infty.$

If the solution of eqns (2.83) and (2.84) does *not* obey the sum rule

$$\frac{1}{\pi}\int_{s_2}^{s_1} \mathrm{d}s' \operatorname{im} F(s'+)D(s') = 0 \tag{2.85}$$

then $\qquad \operatorname{im} D(s+) \sim s^{\frac{1}{2}}, \quad \text{as } s \to +\infty.$

Also $\qquad \operatorname{re} D(s) = O(1), \quad \text{as } s \to +\infty;$

hence $\qquad |D(s+)| \sim s^{\frac{1}{2}}, \quad \text{as } s \to +\infty.$

Thus, by eqn (2.70),

$$\frac{\alpha(\infty)}{\pi} + n_B = \tfrac{1}{2}.$$

However, if the solution to eqns (2.83) and (2.84) obeys the sum rule (2.85), but does *not* obey the sum rule

$$1 + \frac{1}{\pi}\int_{s_0}^{\infty} \frac{q'^3 N(s')}{s'-s_0}\, ds' = 0, \tag{2.86}$$

then, by eqn (2.70),

$$\frac{\alpha(\infty)}{\pi} + n_B = 0.$$

The sum rules (2.85) and (2.86) correspond to restrictions on the left-hand cut discontinuity $\operatorname{im} F(s+)$.

In a similar way we can show that the isolated elastic solution $f_l(s)$ for any angular momentum $l \geqslant 1$ corresponding to a truncated left-hand cut gives

$$\frac{\alpha(\infty)}{\pi} + n_B = \tfrac{1}{2}l,$$

unless a certain sum rule is obeyed.

Let $f_l'(s)$ be a p.w.a. derived from a potential obeying eqns (2.49). Then

$$\frac{\alpha(\infty)}{\pi} + n_B = 0$$

for all l. This shows that if $f_l'(s)$ has a truncated left-hand cut, or a left-hand cut on which $|\operatorname{im} f_l'(s+)|$ vanishes quicker than $|s|^{-\frac{1}{2}}$ as $s \to -\infty$, and $l > 0$, then $f_l'(s)$ obeys a sum rule. In fact we can show that $f_l'(s)$ obeys l sum rules.

Thus for $l > 0$ the set of truncated left-hand cut discontinuities of p.w.a. derived from potentials obeying eqns (2.49) (such as the Bargmann potentials [11]) is a proper subset of the truncated left-hand cuts for which isolated solutions exist.

2.7. The arbitrariness problem for partial wave projections $A(s)$

We now examine briefly the arbitrariness problem for partial wave projections

$$A(s) = s^{\frac{1}{2}} f(s),$$

where $f(s)$ is the usual p.w.a. We saw in Chapter 1 that in π–π scattering for example, it would be natural to specify

$$\rho(s) = \operatorname{im} A(s+)$$

on the left-hand cut. There are two main differences from the arbitrariness problem for $f(s)$ itself which we have so far discussed; first, the extra factor $s^{\frac{1}{2}}$ alters the counting of powers of s as $s \to +\infty$, and, secondly, in many problems the phase shift approaches its limiting value for $s \to +\infty$ like $(\ln s)^{-1}$ (see Chapter 5 for examples of this behaviour).

We shall discuss the set $\mathscr{P}$ of polynomially bounded partial wave projections $A(s)$ which have a given left-hand cut discontinuity $\rho(s)$, and a given inelasticity coefficient $\eta(s)$ on the physical cut $s_0 \leqslant s \leqslant \infty$. We shall assume that in the set $\mathscr{P}$ the equivalent elastic phases $\tilde{\alpha}(s)$ (defined below) obey the Lipschitz condition of eqn (2.10), and in place of eqn (2.11) we shall assume $\tilde{\alpha}(s)$ is of the form

$$\tilde{\alpha}(s) = \tilde{\alpha}(\infty)+\frac{\pi H}{\ln s}+\beta(s), \quad \text{for } s_0 \leqslant s \leqslant \infty, \tag{2.87}$$

where $\tilde{\alpha}(\infty)$ and H are constants and $\beta(s)$ is such that

$$\lim_{s\to+\infty} \beta(s)\ln s = 0$$

and

$$\int\limits_{s_0}^{\infty} \frac{\beta(s')}{s'}\,ds'$$

converges.

By Theorems A and B of Appendix I, these conditions imply that as $s \to +\infty$,

$$\exp\left\{\frac{s}{\pi} P \int\limits_{s_0}^{\infty} \frac{\tilde{\alpha}(s')\,ds'}{s'(s'-s)}\right\} \to C(\ln s)^{-H} s^{-\tilde{\alpha}(\infty)/\pi}, \tag{2.88}$$

where C is a positive constant.

The relation

$$2iqs^{-\frac{1}{2}}\tilde{A}(s)+1 = L(s)\{2iqs^{-\frac{1}{2}}A(s)+1\} \tag{2.89}$$

enables us to construct the equivalent elastic amplitude $\tilde{A}(s)$. Here $s^{\frac{1}{2}}$ is defined as in Fig. 1.1 and $L(s)$ is Froissart's function (eqn (2.20)). By eqn (2.21)

$$\tilde{A}(s+) = s^{\frac{1}{2}}\,\frac{\exp(2i\tilde{\alpha}(s))-1}{2iq}, \quad \text{for } s_0 \leqslant s \leqslant \infty, \tag{2.90}$$

where

$$\tilde{\alpha}(s) = \alpha(s)+\tfrac{1}{2}\phi(s).$$

The left-hand cut discontinuity of $\tilde{A}(s)$ is given by

$$\begin{aligned}\tilde{\rho}(s) &\equiv \operatorname{im} \tilde{A}(s+)\\ &= L(s)\rho(s)+\operatorname{im}\left\{\frac{s^{\frac{1}{2}}}{2iq}\,(L(s)-1)\right\}.\end{aligned} \tag{2.91}$$

Thus $\tilde{\rho}(s)$ is determined by $\rho(s)$ and $\eta(s)$. In the case of π–π scattering $s_0 = 4m_\pi^2$ and the left-hand cut is $-\infty \leqslant s \leqslant 0$. On the line $-\infty \leqslant s < s_0$, $L(s)$ is real and $q(s)$ is regular, and $q = i|q|$. Thus the second term on the right of eqn (2.91) will give an extra contribution to $\tilde{\rho}(s)$ on $-\infty \leqslant s < 0$; this extra contribution depends on $\eta(s)$ only.

We assume, as in § 2.3 above, that the function $\eta(s)$ is such that $L(s)$ is polynomially bounded.

The spectrum of values of $\tilde{\alpha}(\infty)$ *and* H

Let $A_1(s)$ and $A_2(s)$ be two partial wave projections in $\mathscr{P}$ having n_1 and n_2 bound states with positions B_i $(i = 1, 2,..., n_1)$ and B_i' $(i = 1, 2,..., n_2)$ respectively. We define

$$\Delta\tilde{A}(s) = \tilde{A}_2(s) - \tilde{A}_1(s).$$

Then

$$\Delta\tilde{A}(s+) = \left(\frac{s^{\frac{1}{2}}}{q}\right)\sin(\alpha_2(s)-\alpha_1(s))\exp\{i(\tilde{\alpha}_2(s)+\tilde{\alpha}_1(s))\}, \quad \text{for } s_0 \leqslant s \leqslant \infty. \tag{2.92}$$

By the arguments used in § 2.4 we find that for all s

$$\Delta\tilde{A}(s)\prod_{i=1}^{n_1}(s-B_i)\prod_{i=1}^{n_2}(s-B_i') = (s-s_0)^l E(s)\,\tilde{A}_{12}(s), \tag{2.93}$$

where l is the orbital angular momentum, $E(s)$ is a polynomial of degree $(n-1)$, and $\tilde{A}_{12}(s)$ is given by eqn (2.34).

It follows from eqns (2.92) and (2.93) that as $s \to +\infty$,

$$|\sin(\alpha_2(s)-\alpha_1(s))|\prod_{i=1}^{n_1}(s-B_i)\prod_{i=1}^{n_2}(s-B_i')$$
$$\to qs^{-\frac{1}{2}}(s-s_0)^l|E(s)|\tilde{C}(\ln s)^{-(H_1+H_2)}s^{-(1/\pi)\{\tilde{\alpha}_1(\infty)+\tilde{\alpha}_2(\infty)\}},$$

where $\tilde{C}$ is a positive constant. This gives us the basic results. We write

$$\psi \equiv n-1+l-\frac{1}{\pi}(\tilde{\alpha}_1(\infty)+\tilde{\alpha}_2(\infty))-(n_1+n_2).$$

If $\tilde{\alpha}_2(\infty)-\tilde{\alpha}_1(\infty) \neq 0 \pmod{\pi}$,

then $$\psi = 0 \quad \text{and} \quad H_1+H_2 = 0. \tag{2.94}$$

If $\tilde{\alpha}_2(\infty)-\tilde{\alpha}_1(\infty) = 0 \pmod{\pi}$,

then $$\psi = 0 \quad \text{and} \quad H_1+H_2 = 1, \quad \text{if } H_1 \neq H_2; \tag{2.95}$$

$$\psi \leqslant 0, \quad \text{if } H_1 = H_2 > \tfrac{1}{2}; \tag{2.96}$$

$$\psi < 0, \quad \text{if } H_1 = H_2 \leqslant \tfrac{1}{2}. \tag{2.97}$$

So we have the following lemma.

Lemma 5. *If A_1 and A_2 are two distinct partial wave projections in $\mathscr{P}$ then either*

$$\tilde{\alpha}_2(\infty)-\tilde{\alpha}_1(\infty) = 0 \pmod{\pi}$$

or

$$\tilde{\alpha}_2(\infty)+\tilde{\alpha}_1(\infty) = 0 \pmod{\pi}.$$

Also

$$\frac{\tilde{\alpha}_1(\infty)}{\pi}+n_1+\frac{\tilde{\alpha}_2(\infty)}{\pi}+n_2 \geqslant l.$$

Let A be a solution in $\mathscr{P}$ with $\tilde{\alpha}(\infty) = a$, and $H = h$. By Lemma 5 any solution in $\mathscr{P}$ has

$$\tilde{\alpha}(\infty) = a \pmod{\pi}, \tag{2.98}$$

or

$$\tilde{\alpha}(\infty) = b \pmod{\pi}, \tag{2.99}$$

where $b = -a \pmod{\pi}$, and $a \leqslant b < a+\pi$.

Thus the spectrum of possible values of $\tilde{\alpha}(\infty)$ is of the same form as the spectrum in Fig. 2.3. Solutions obeying eqn (2.98) we shall call *type a* and solutions obeying eqn (2.99) we shall call *type b*.

The results in eqns (2.94), (2.95), (2.96), and (2.97) on the values of H can be summarized by a further lemma.

Lemma 6. *If $a = b$ or if there are no solutions of type b, then any solution must have $H = h$ or $H = 1-h$. If $a \neq b$ and if there exist solutions of both types, then solutions of the same type all have the same value of H, and solutions of different type have values of H which have the same magnitude but opposite signs. In each case there are at most two distinct values of H.*

Let $A(s)$ be a solution in $\mathscr{P}$ having n_B bound states and the equivalent elastic phase $\tilde{\alpha}(s)$. The *index* of $A(s)$ is the integer p given by

$$p \leqslant 2\left(\frac{\tilde{\alpha}(\infty)}{\pi}+n_B\right)-l+1 < p+1.$$

It follows from the last part of Lemma 5 that there is at most one solution with index $p \leqslant 0$. We call this the *isolated solution.*

Suppose $A_1(s)$ is a solution with index $p_1 = 1$, $H_1 \leqslant \frac{1}{2}$, and $2\tilde{\alpha}_1(\infty) = 0 \pmod{\pi}$. Then

$$\frac{\tilde{\alpha}_1(\infty)}{\pi}+n_1 = \tfrac{1}{2}l \tag{2.100}$$

Let $A_2(s)$ be a solution distinct from $A_1(s)$. By eqn (2.100) and the last part of Lemma 5, we must have either

$$\frac{\tilde{\alpha}_2(\infty)}{\pi}+n_2 > \tfrac{1}{2}l, \tag{2.101}$$

or

$$\frac{\tilde{\alpha}_2(\infty)}{\pi}+n_2 = \tfrac{1}{2}l. \tag{2.102}$$

In the case of eqn (2.102) we have $\psi = n-1$ in eqns (2.94), (2.95), (2.96), and (2.97), so $\psi \geqslant 0$, since $(n-1)$ is the degree of the polynomial $E(s)$ in eqn (2.93). Thus by eqns (2.94), (2.95), (2.96), and (2.97) we must have $H_2 = 1-H_1$ and $H_2 > \frac{1}{2}$. Therefore any solution distinct from $A_1(s)$ must either have a larger index (which is the case when eqn (2.101) holds) or else have $H = 1-H_1$ and $H > \frac{1}{2}$. This gives the result.

THEOREM 8. *There is at most one solution $A(s)$ in $\mathscr{P}$ with index $p \leqslant 0$ (the isolated solution). There is at most one solution $A(s)$ in $\mathscr{P}$ with index $p = 1$, $H = h \leqslant \frac{1}{2}$, and $2\tilde{\alpha}(\infty) = 0 \pmod{\pi}$. If such a solution exists, any other solution will either have a larger index, or else will have $H = 1-h$, $H > \frac{1}{2}$.*

The methods of Appendix II are easily modified to solve eqn (2.93), i.e. we can determine $A_2(s)$ when $A_1(s)$ and $E(s)$ are given. We quote some of the results on the structure of the sets of solutions. The metric to be used in $\mathscr{P}$ is

$$d(A_1, A_2) = \sup_{s_0 \leqslant s \leqslant \infty} |\alpha_1(s)-\alpha_2(s)|.$$

THEOREM 9. *Let $A(s)$ be a solution in the set $\mathscr{P}$ having n bound states, index p, and $H = h$. Then*

(*a*) *if $p > 0$, $A(s)$ is contained in a manifold of solutions having the same index p. If $h \leqslant \frac{1}{2}$ and $2\tilde{\alpha}(\infty) = 0 \pmod{\pi}$, the dimension of the manifold is p or $(p-1)$. In all other cases the dimension is p.*

(*b*) *There is a neighbourhood of $A(s)$ which contains no solutions of index less than p.*

(*c*) *Let m be an integer such that $m \geqslant n$ and $p+m-n > 0$. In each neighbourhood of $A(s)$ there is a manifold of solutions having m bound states and index $p_m = p+2(m-n)$. The dimension of the manifold is p_m or (p_m-1), following the rules in (a).*

A p.w.a. $f'(s)$ derived from a potential obeying eqns (2.49) obeys Levinson's theorem, i.e.

$$\frac{\alpha(\infty)}{\pi}+n = 0.$$

Thus $A'(s) = s^{\frac{1}{2}}f'(s)$ has index $(1-l)$. If $l > 0$ there are no other partial wave projections having the same left-hand cut discontinuity and the same index. If $l = 0$ there may be a one-dimensional manifold of such partial wave projections.

3

GENERAL PROPERTIES OF SECOND SHEET POLES OF ELASTIC PARTIAL WAVE AMPLITUDES

3.1. Properties of the phase $\alpha(s)$

HERE we restrict the discussion to elastic amplitudes, which may, however, have any (integral) orbital angular momentum l. The kinematics in an unequal mass case, such as $\pi N \to \pi N$, are given by

$$q^2 = \frac{(s-s_0)(s-s_1)}{4s}, \tag{3.1}$$

with $s_0 = (M+\mu)^2$ and $s_1 = (M-\mu)^2$, M and μ being the masses of the two particles. For equal masses, $s_1 = 0$ and

$$q^2 = \tfrac{1}{4}(s-s_0). \tag{3.2}$$

In either case we define $q(s)$ in the s-plane cut along $s_0 \leqslant s \leqslant \infty$ (Fig. 3.1).

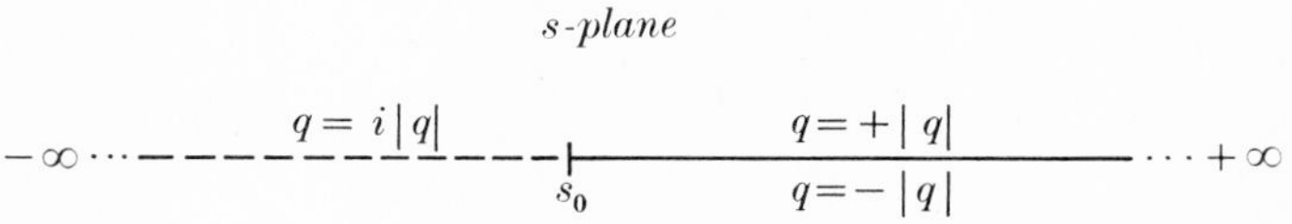

s-plane

$-\infty \cdots$ - - - $q = i|q|$ - - - s_0 ——— $q = +|q|$ / $q = -|q|$ ——— $\cdots +\infty$

FIG. 3.1. The cut $s_0 \leqslant s \leqslant \infty$ in the definition of $q(s)$.

For unequal masses it is also cut along $0 \leqslant s \leqslant s_1$. Thus near the physical cut

$$q(s \pm i\epsilon) = \pm|q(s)|, \quad \text{for } s_0 \leqslant s \leqslant \infty, \quad \epsilon > 0.$$

In general

$$q^*(s) = -q(s^*). \tag{3.3}$$

The phase shift $\alpha(s)$ is an analytic function of s defined by

$$f(s) = \frac{\exp(2i\alpha(s))-1}{2iq(s)}, \tag{3.4}$$

where $f(s)$ is the partial wave amplitude. We take $\alpha(s_0) = 0$ at the start of the physical cut. Since

$$\alpha(s) = \frac{1}{2i}\ln(1+2iq(s)f(s)), \tag{3.5}$$

the phase $\alpha(s)$ is singular at the poles and zeros of $(1+2iq(s)f(s))$. In addition $\alpha(s)$ has the cuts of $f(s)$. We can, if we wish, draw branch lines from the zeros and poles of $(1+2iq(s)f(s))$ to infinity in order to obtain a single-valued definition of $\alpha(s)$. This will sometimes be done.

The reality condition $$f(s^*) = f^*(s)$$
implies $$\exp(-2i\alpha^*(s)) = \exp(2i\alpha(s^*)),$$
so $$\alpha(s^*) = -\alpha^*(s) \pmod{\pi} \tag{3.6}$$
on the physical s-plane. If cuts are put in to make $\alpha(s)$ single-valued, then using $\alpha(s_0) = 0$ we have
$$\alpha(s^*) = -\alpha^*(s). \tag{3.7}$$
It follows from eqn (3.3) that $\alpha(s)$ must be an odd function of $q(s)$ near the physical threshold s_0.

As s approaches the physical cut $s_0 \leqslant s \leqslant \infty$, $\alpha(s)$ becomes real because of elastic unitarity, and by eqn (3.7)
$$\alpha(s+i\epsilon) = -\alpha(s-i\epsilon) \quad \text{for } s_0 \leqslant s \leqslant \infty. \tag{3.8}$$

The function $S(s)$

We define $$\left.\begin{aligned} S(s) &= 1+2iq(s)f(s) \\ &= \exp(2i\alpha(s)) \end{aligned}\right\}. \tag{3.9}$$
$S(s)$ is an analytic function having the same cuts and poles as $f(s)$. On the physical cut $|S(s)| = 1$. By eqn (3.8) (or (3.6)), for $s_0 \leqslant s \leqslant \infty$,
$$\begin{aligned} (S(s+i\epsilon))^{-1} &= \exp(-2i\alpha(s+i\epsilon)) \\ &= \exp(2i\alpha(s-i\epsilon)) \\ &= S(s-i\epsilon). \end{aligned} \tag{3.10}$$

There is a theorem, discussed, for example, in § 4.5 of Titchmarsh [16]: *if two functions $f(z)$, $f_1(z)$ are regular in regions D and D_1 which are separated by a contour C, and if they are continuous on C and $f(z) = f_1(z)$ on C, then the two functions are analytic continuations of each other.* Using eqn (3.10) this theorem shows that on continuing $S(s)$ through the physical cut $s_0 \leqslant s \leqslant \infty$, we get (see Fig. 3.2)
$$S^{\mathrm{II}}(s) = (S(s))^{-1}. \tag{3.11}$$
The notation $S^{\mathrm{II}}(s)$ is used to indicate the value of the function on the second, or lower, sheet.

Continuing $q(s)$ through $s_0 \leqslant s \leqslant \infty$ gives $q^{\mathrm{II}}(s) = -q(s)$, and likewise continuing $f(s)$ gives the second sheet function $f^{\mathrm{II}}(s)$. So by eqn (3.9) we have
$$S^{\mathrm{II}}(s) = 1-2iq(s)f^{\mathrm{II}}(s) = 1/(1+2iq(s)f(s)),$$
or $$f^{\mathrm{II}}(s) = \frac{f(s)}{1+2iq(s)f(s)}. \tag{3.12}$$

This continuation is only possible because of unitarity. Equation (3.12) has been well known for many years [20].

The poles of $f^{\text{II}}(s)$ (i.e. the second sheet poles of $f(s)$) are associated with resonances. It is more convenient to work with the physical sheet function $S(s)$. Poles of $S(s)$ give the poles of $f(s)$ (eqn (3.9)), but the zeros of $S(s)$ give the poles of $f^{\text{II}}(s)$ (eqn (3.11)).

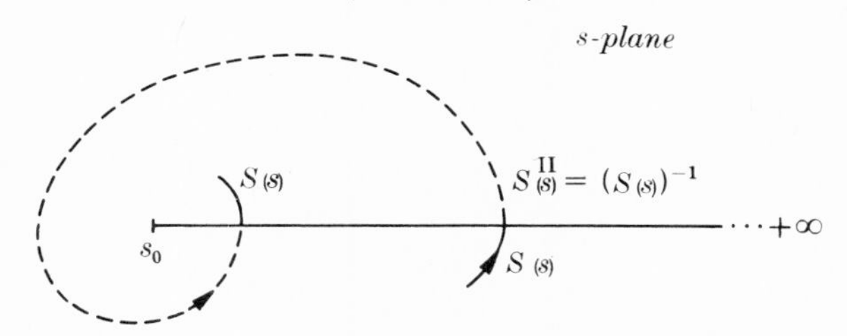

FIG. 3.2. Continuation of $S(s)$ through the physical cut $s_0 \leqslant s \leqslant \infty$.

At a pole of $f(s)$, $\text{im}\,\alpha(s) \to -\infty$, whereas at a pole of $f^{\text{II}}(s)$, $\text{im}\,\alpha(s) \to +\infty$. Curves $\text{im}\,\alpha(s) = 0$ must pass between these two kinds of poles; they are branches of the curve $|S| = 1$.

3.2. Physical phases and the unitarity curve

The locus obeying $|S(s)| = 1$ will be called the unitarity curve. We use the following simple result of function theory: except for singular points (i.e. poles and branch points) the branches of the unitarity curve $|S| = 1$ can only meet where $\mathrm{d}S/\mathrm{d}s = 0$ (and $|S| = 1$). If $\mathrm{d}S/\mathrm{d}s$ has a simple zero, two branches meet orthogonally; if $\mathrm{d}S/\mathrm{d}s$ has a zero of order n, then $(n+1)$ branches meet with angular separation $\pi/(n+1)$.

The physical axis is always a branch of the unitarity curve. The result above shows that other branches of the unitarity curve can only meet the physical axis at the threshold s_0 and at points where

$$\frac{\mathrm{d}S(s+i\epsilon)}{\mathrm{d}s} = 0, \quad \text{i.e. where } \frac{\mathrm{d}\alpha(s+i\epsilon)}{\mathrm{d}s} = 0.$$

One or several branches of the unitarity curve will meet the physical axis at each point where $\mathrm{d}\alpha(s)/\mathrm{d}s = 0$. Where $\alpha(s+i\epsilon)$ has a maximum or a minimum, $\mathrm{d}\alpha(s)/\mathrm{d}s = 0$ and $\mathrm{d}^2\alpha/\mathrm{d}s^2 \neq 0$, so that one branch of the unitarity curve meets the physical axis at such a point, and crosses the physical axis orthogonally.

Let $\alpha(s)$ be regular at $s = (x, y)$. At a point $s' = (x', y')$ near s we can expand $\alpha(s')$ as

$$\alpha(s') = \alpha(s) + (s'-s)\frac{\mathrm{d}\alpha(s)}{\mathrm{d}s} + \tfrac{1}{2}(s'-s)^2\frac{\mathrm{d}^2\alpha(s)}{\mathrm{d}s^2} + \dots. \tag{3.13}$$

Put s on the physical cut† (but not at s_0). On that cut $\alpha(s)$ and all its derivatives are real. By eqn (3.13)

$$\operatorname{im}\alpha(s') = y'\,\frac{d\alpha(s+i\epsilon)}{ds} + y'(x'-x)\,\frac{d^2\alpha(s+i\epsilon)}{ds^2} + (\text{3rd order terms}). \tag{3.14}$$

The first term on the right of eqn (3.14) shows that $\operatorname{im}\alpha > 0$, i.e. $|S| < 1$, just above any segment of the physical axis where the phase shift is increasing with energy; similarly, $\operatorname{im}\alpha < 0$, i.e. $|S| > 1$, just above any segment where the phase shift is decreasing with energy. It also shows the property mentioned above, that a single branch of the unitarity curve meets the physical axis orthogonally at any maximum or minimum of the phase shift.

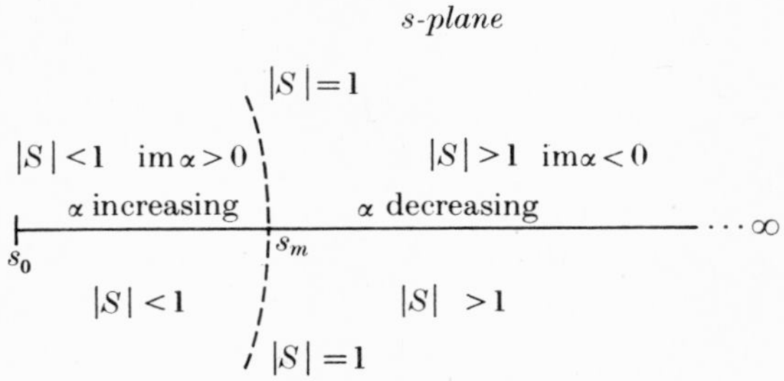

FIG. 3.3. An example of the behaviour of $|S(s)|$ near the physical cut. There is a maximum of $\alpha(s)$ at s_m.

We use eqn (3.13) on the straight line $\operatorname{re} s' = s_m$, $\operatorname{im} s' \geqslant 0$, where $\alpha(s)$ has a maximum or minimum at the point s_m on the physical axis. Then

$$\operatorname{re}\alpha(s') = \alpha(s_m+i\epsilon) - \tfrac{1}{2}(\operatorname{im} s')^2\,\frac{d^2\alpha(s_m+i\epsilon)}{ds^2} + O((\operatorname{im} s')^4), \tag{3.15}$$

$$\operatorname{im}\alpha(s') = -\tfrac{1}{6}(\operatorname{im} s')^3\,\frac{d^3\alpha(s_m+i\epsilon)}{ds^3} + O((\operatorname{im} s')^5). \tag{3.16}$$

It follows from eqn (3.15) that if $\alpha(s+i\epsilon)$ has a maximum or minimum at s_m then $\alpha(s)$ respectively increases or decreases as we move into the region $\operatorname{im} s > 0$ along the branch $|S| = 1$ (remember that $\alpha(s)$ is real on any branch of the unitarity curve). From eqn (3.16) it follows that in the situation shown in Fig. 3.3, the curve $|S| = 1$ will bend off to the left of the line $\operatorname{re} s' = s_m$ if $\dfrac{d^3}{ds^3}\alpha(s_m+i\epsilon) > 0$, and to the right of this line if $\dfrac{d^3}{ds^3}\alpha(s_m+i\epsilon) < 0$, as $\operatorname{im} s'$ increases from zero.

† We include a small region of the second sheet so as to give a suitable domain of regularity for $\alpha(s)$ at s.

A resonance pole, i.e. a pole of $f^{\mathrm{II}}(s)$, requires $S(s) = 0$. It can only lie within regions $|S| \leqslant 1$ bounded by the unitarity curve. In particular there cannot be a resonance pole near a segment of the physical cut where $\alpha(s+i\epsilon)$ is decreasing; there cannot be a resonance pole associated with $\alpha(s+i\epsilon)$ decreasing through $\frac{1}{2}\pi$.

3.3. Simple properties of the curves re S = constant, im S = constant, and $|S| = 1$

We write

$$S(s) = R(s)+iI(s)$$

where R and I are real (harmonic) functions. The family of curves

$$R(x,y) = \text{constant}$$

is orthogonal to the family

$$I(x,y) = \text{constant},$$

except at singular points of $S(s)$ and at zeros of $\mathrm{d}S/\mathrm{d}s$. Excluding singular points there is one curve $R = c$ and one curve $I = c'$ through each point where $\mathrm{d}S/\mathrm{d}s \neq 0$. At a simple zero of $\mathrm{d}S/\mathrm{d}s$ two curves $R = c$ meet orthogonally, and two curves $I = c'$ meet orthogonally, the angle between the two sets being $\frac{1}{4}\pi$ (see Fig. 3.4).

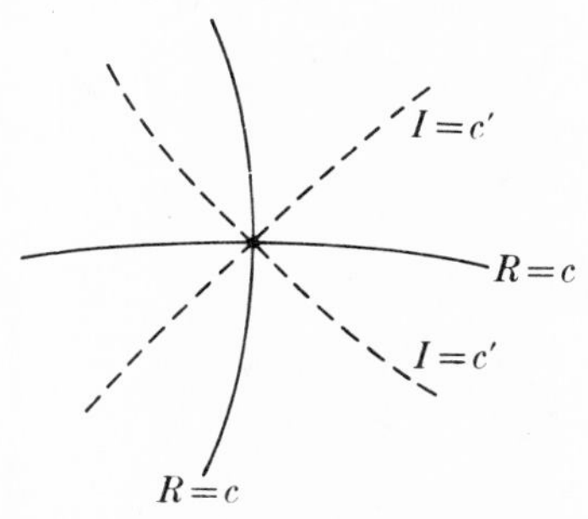

FIG. 3.4. Intersection of the level curves $R = c$, $I = c'$ at a simple zero of $\mathrm{d}S/\mathrm{d}s$.

If we move along the curve $R = c$ so that the region $R > c$ is on the right, then I increases (Fig. 3.5 (a)). If we move along the curve $I = c'$ so that the region $I > c'$ is on the right, then R is decreasing (Fig. 3.5 (b)). These results come from using the Cauchy–Riemann relations in the form

$$\frac{\partial R}{\partial n} = \frac{\partial I}{\partial t}, \qquad \frac{\partial I}{\partial n} = -\frac{\partial R}{\partial t},$$

where $\mathbf{n}$ and $\mathbf{t}$ are orthogonal directions and $\mathbf{n}\times\mathbf{t}$ points up out of the complex plane.

On any branch of $|S| = 1$, $\alpha(s)$ is real. Let t measure the distance along $|S| = 1$ on any branch of $|S| = 1$ except the physical cut. We can write

$$R(t) = \cos(2\alpha(t)), \qquad I(t) = \sin(2\alpha(t)), \tag{3.17}$$

and

$$\frac{dS}{dt} = 2i\exp(2i\alpha)\frac{d\alpha}{dt}.$$

Therefore $d\alpha/dt = 0$ implies $dS/ds = 0$ except at a singular point of S. This gives us a further theorem.

THEOREM 10. *The phase $\alpha(t)$ is monotonic increasing, or monotonic decreasing, on any segment of the curve $|S| = 1$ which has no double points and no singular points.*† *The same is true for R (I), except at points where $R = \pm 1$ $(I = \pm 1)$.*

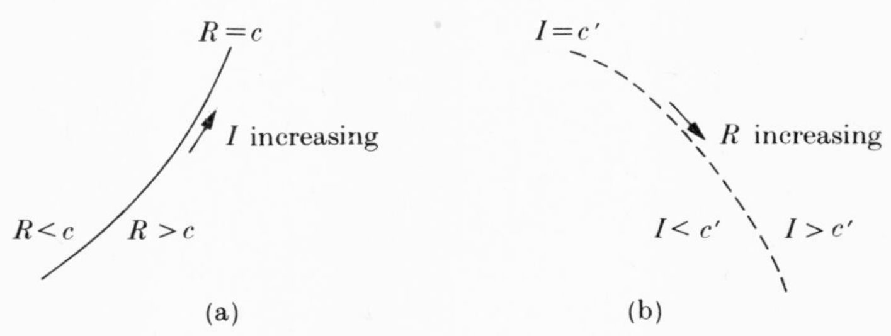

FIG. 3.5. (a) The direction along R = constant in which I increases. (b) The direction along I = constant in which R increases.

The last part of the theorem follows from eqn (3.17). It means that a curve $R = c$ cannot be tangential to $|S| = 1$ away from a double point or a singular point unless $c = \pm 1$. The same is true for a curve $I = c'$ unless $c' = \pm 1$.

Suppose dS/ds has a zero of order n at $\bar{s}$ and $|S(\bar{s})| = 1$. Let β be the argument of $d^{n+1}\alpha(\bar{s})/ds^{n+1}$. Then near $\bar{s}$,

$$\alpha(s) = \alpha(\bar{s}) + \frac{|s-\bar{s}|^{n+1}}{(n+1)!}\left|\frac{d^{n+1}\alpha(\bar{s})}{ds^{n+1}}\right|\exp\{i(\beta+(n+1)\theta)\} + O(|s-\bar{s}|^{n+2}), \tag{3.18}$$

where $\theta = \arg(s-\bar{s})$. There are $(n+1)$ branches C_i of the unitarity curve passing through $\bar{s}$. The angle θ_i of the tangent to C_i at $\bar{s}$ obeys

$$\beta+(n+1)\theta_i = 0 \pmod{\pi}.$$

Also

$$\beta+(n+1)\theta_i$$

increases, or decreases, by π when we go from one branch C_i to an adjacent branch $C_{(i+1)}$. We define the sense of the tangents by the

† An nth order zero of dS/ds will be called a double point even if $n > 1$.

condition $|\theta_{i+1}-\theta_i| \leqslant \frac{1}{2}\pi$, for $n \geqslant 1$. Equation (3.18) shows that if $\alpha(s)$ increases or decreases as we move from $\bar{s}$ along a branch C_i, then $\alpha(s)$ will respectively decrease or increase as we move from $\bar{s}$ along an adjacent branch $C_{(i+1)}$. For the cases $n = 1$ and $n = 2$ this gives the behaviour shown in Figs. 3.6 (a) and 3.6 (b).

A simple example is the behaviour of $\alpha(s)$ on the branch of the unitarity curve in $\operatorname{im} s > 0$ which meets the physical cut where $\alpha(s+i\epsilon)$ has a maximum or a minimum. In general if $\alpha(s)$ is increasing as we move towards a double (or multiple) point $\bar{s}$, then $\alpha(s)$ will again increase if

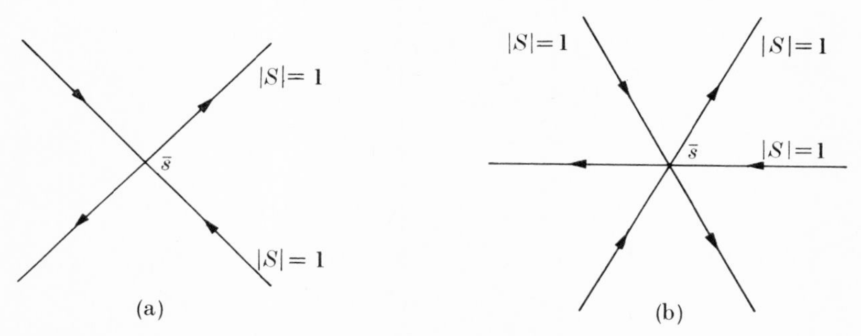

FIG. 3.6. (a) Branches of $|S(s)| = 1$ at a double point $\bar{s}$. The arrows show the directions of increasing $\alpha(t)$. (b) Branches of $|S(s)| = 1$ at a triple point $\bar{s}$. The arrows show the directions of increasing $\alpha(t)$.

we move away from $\bar{s}$ along the next branch, either to the right or the left.

Suppose we can form a closed circuit out of branches of $|S| = 1$ without crossing a cut of $S(s)$. On moving once around this circuit in a counter-clockwise direction, the variation $\Delta\alpha$ of α will be

$$\Delta\alpha = \pi(N_0 - N_p), \tag{3.19}$$

where N_0, N_p are the number of zeros and poles of $S(s)$ which lie inside the closed circuit.

We could if we wished arrange cuts of $\alpha(s)$ so as to keep $\alpha(s)$ single-valued. The modification required in eqn (3.19) is simple; we replace $\Delta\alpha$ by minus the sum of the discontinuities in $\alpha(s)$ across the cuts traversed by the circuit.

3.4. Use of the maximum modulus principle

The maximum modulus theorem, discussed in Titchmarsh [16], Chapter 5, states that if a function $g(z)$ is regular in a region D, then $|g(z)|$ cannot have a maximum *inside* D. If $|g(z)|$ has a minimum inside D, the minimum must be zero.

Thus a region which is entirely bounded by branches of the unitarity curve $|S| = 1$ must *either* contain at least one (physical sheet) singularity of $S(s)$, *or else* must contain at least one zero of $S(s)$. For example, the situation in Fig. 3.7 is forbidden.

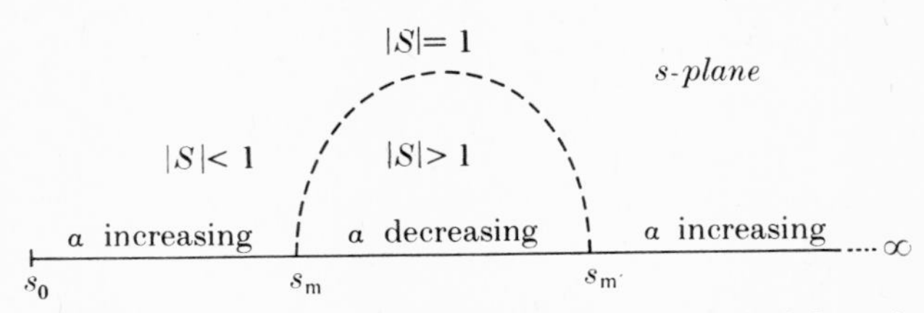

FIG. 3.7. This configuration is impossible. It implies a physical sheet singularity of $S(s)$ within the region bounded by the unitarity curve.

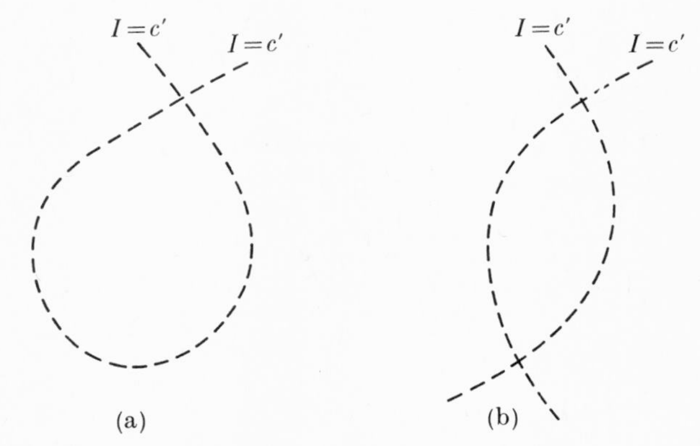

FIG. 3.8. (a), (b) The regions enclosed by level curves $I = c'$ must contain a singularity of $S(s)$.

Applying the maximum modulus principle to $\exp(S(s))$ or $\exp(iS(s))$ it follows that R and I cannot have a maximum or a minimum *inside* any region where $S(s)$ is regular. An interesting example is shown in Fig. 3.8. We have regions enclosed by a curve, or curves, $I =$ constant. At points inside (ignoring the trivial case $I(s) =$ constant, i.e. $S(s) =$ constant), I must exceed, or be less than, this value. But I cannot have a maximum or minimum where $S(s)$ is regular, therefore the enclosed regions must contain a singularity of $S(s)$. A similar result holds for R.

Another example is shown in Fig. 3.9. A region is enclosed by the curves $R = c$ and $I = c'$. Without loss of generality (we merely replace

$S(s)$ by $S(s)-c-ic'$), we can label the curves $R = 0$, $I = 0$ respectively. Now

$$IR = \tfrac{1}{2}\operatorname{im}(S^2(s)),$$

therefore $\operatorname{im}(S^2) = 0$ on the boundary of the region in Fig. 3.9. The preceding example shows that $S(s)$ must have a singularity in the enclosed region.

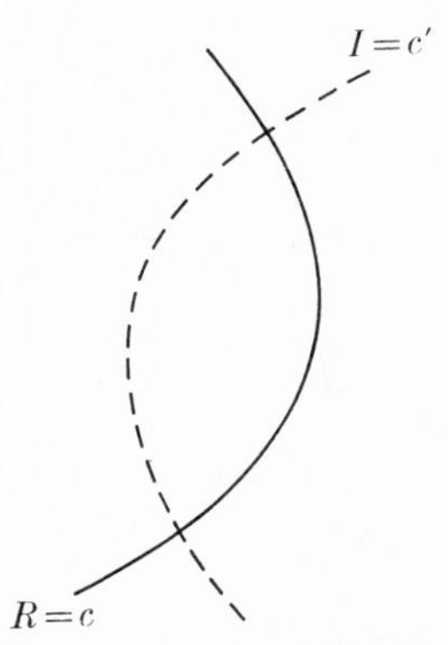

FIG. 3.9. A region enclosed by the level curves $R = c$, $I = c'$ must contain a singularity of $S(s)$.

3.5. Behaviour near the physical threshold

We first examine the function $q(s)$ in more detail. In the case of unequal masses eqn (3.1) holds. For a point s (on the physical sheet) we use the notation in Fig. 3.10. Thus

$$\left.\begin{aligned} q(s) &= \frac{1}{2}\left(\frac{r_0 r_1}{r}\right)^{\frac{1}{2}} \exp\{\tfrac{1}{2}i(\theta_0+\theta_1-\theta)\} \\ &= |q(s)|\exp(i\phi) \end{aligned}\right\}. \tag{3.20}$$

Typical values of ϕ in the unequal mass case are shown in Fig. 3.11.

In the equal mass case

$$\phi = \tfrac{1}{2}\theta_0 = \tfrac{1}{2}\arg(s-s_0) \tag{3.21}$$

(see Fig. 3.12). In the equal mass and unequal mass cases $|q(s)|$ behaves like $|s-s_0|^{\frac{1}{2}}$ near the physical threshold s_0.

We saw in § 3.1 that $\alpha(s)$ has to be an odd function of $q(s)$ near s_0. For orbital angular momentum l we have

$$\alpha(s) = (q(s))^{2l+1}\{a+b(q(s))^2+...\}, \tag{3.22}$$

where a, b,... are constants. We call a the scattering length.

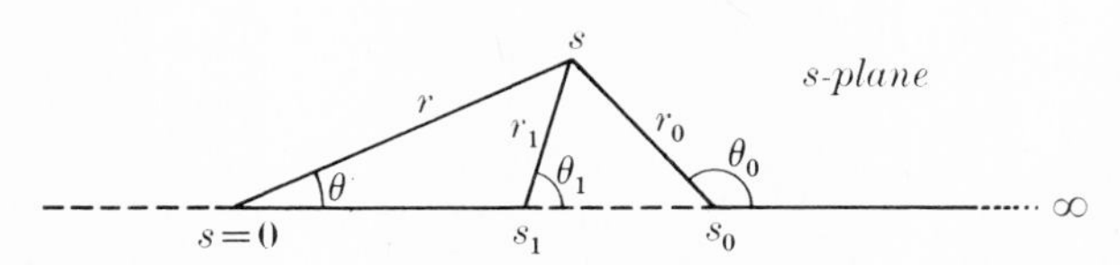

FIG. 3.10. Notation used for $q(s)$ in the unequal mass case. The plane is cut along $0 \leqslant s \leqslant s_1$ and $s_0 \leqslant s \leqslant \infty$.

We first look at the form of the curve $|S| = 1$ near s_0. Since

$$\frac{dS}{dq} = 2i\frac{d\alpha}{dq}S,$$

it follows from eqn (3.22), assuming $a \neq 0$, that at s_0, dS/dq has a zero of order $2l$ in q. A result from function theory, similar to that given at

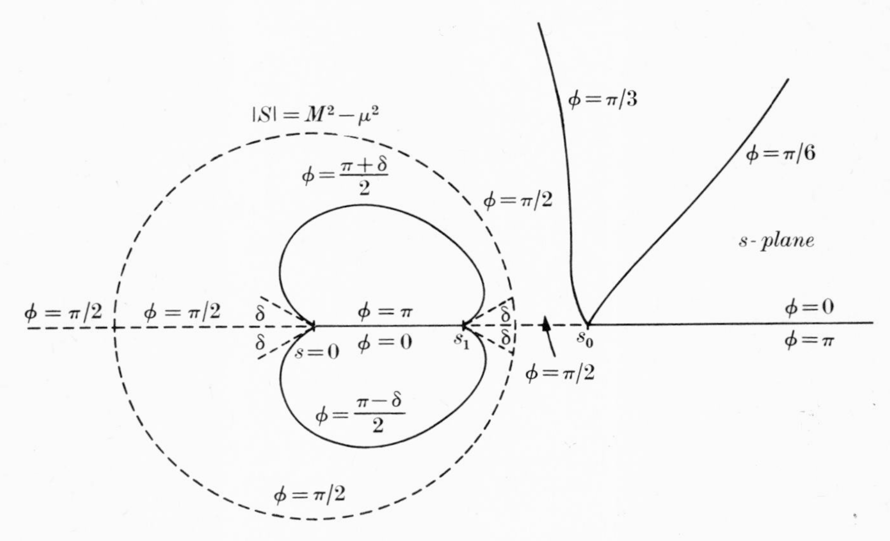

FIG. 3.11. Values of $\phi(s) = \arg q(s)$ in the unequal mass case. ($s_0 = (M+\mu)^2$ $s_1 = (M-\mu)^2$.)

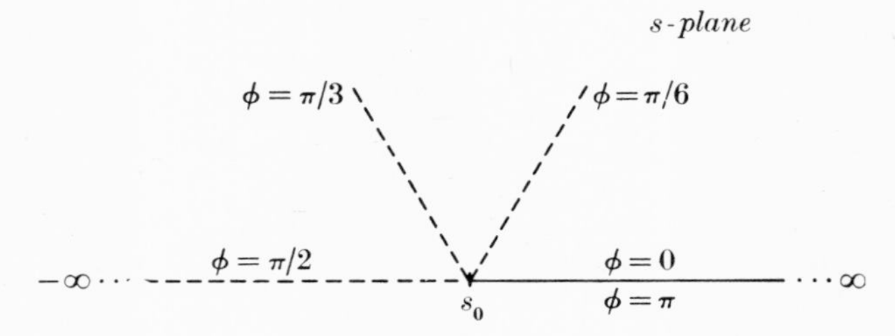

FIG. 3.12. Values of $\phi(s) = \arg q(s)$ in the equal mass case.

the start of § 3.2, shows that $(2l+1)$ branches of $|S| = 1$ pass through $q = 0$ ($s = s_0$) in the q-plane and their angular separation is $\pi/(2l+1)$.

By eqns (3.20) or (3.21), $0 \leqslant \phi \leqslant \pi$ on the physical sheet of the s-plane near s_0. Hence $(2l+1)$ branches of the curve $|S| = 1$ meet at s_0 on the physical sheet. One of them is the physical cut; the other $2l$ branches approach s_0 tangential to the curves

$$\phi = \frac{k\pi}{2l+1}, \quad \text{where } k = 1, 2, \ldots, 2l \text{ and } l \geqslant 1.$$

For $l = 0$ the only branch of $|S| = 1$ reaching s_0 is the physical cut. For $l = 1$ there are two additional branches which approach s_0 tangential to $\phi = \frac{1}{3}\pi$ and $\phi = \frac{2}{3}\pi$ (Fig. 3.13) (except for $a = 0$; in that case the situation is similar to a D-wave having $a \neq 0$).

At s_0, $R = 1$, $I = 0$, and we should find how the curve $I = 0$ reaches s_0. Since

$$S(s) = \exp(2i\,\mathrm{re}\,\alpha)\exp(-2\,\mathrm{im}\,\alpha),$$

the curve $I = 0$ is $\mathrm{re}\,\alpha = 0$ (mod $\frac{1}{2}\pi$). Near s_0, $|\alpha(s)|$ is small, and by eqn (3.22) $\mathrm{re}\,\alpha = 0$ gives

$$a\cos((2l+1)\phi)+b|q(s)|^2\cos((2l+3)\phi)+\ldots = 0.$$

For $a \neq 0$, we require $\cos((2l+1)\phi) = 0$ near s_0, i.e.

$$\phi = \pi/2(2l+1),\ 3\pi/2(2l+1),\ldots .$$

On the physical s-sheet, for an S-wave only $\phi = \frac{1}{2}\pi$ occurs; for a P-wave $\phi = \frac{1}{6}\pi$ and $\phi = \frac{5}{6}\pi$ also give $I = 0$ near s_0 (Fig. 3.14). The reality condition for $f(s)$ shows that in all cases the real axis from s_0 to the first branch point having $s < s_0$ will be a branch of the curve $I = 0$.

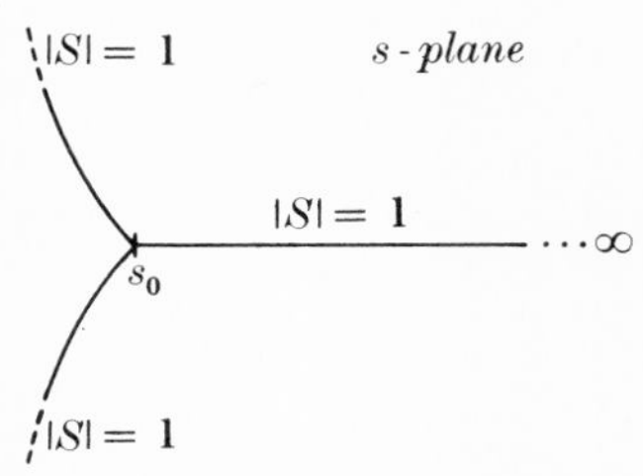

FIG. 3.13. For a P-wave three branches of $|S| = 1$ meet at the threshold s_0 with angular separation $\frac{2}{3}\pi$. (The scattering length is not zero.)

3.6. Example: elastic P-wave with a single attractive driving pole

The general results on unitarity curves are of value in helping us to find where resonance poles can lie. We shall illustrate this by considering the isolated elastic P-wave solution $f(s)$ which has a left-hand cut represented by an attractive driving pole $\Gamma/(s-P)$ in the reduced p.w.a.

$$F(s) = f(s)/q^2.$$

We have $\Gamma > 0$, and $-\infty < P < s_0$. The equal mass case $q^2 = \frac{1}{4}(s-s_0)$ is considered.

This P-wave is the solution of the form

$$F(s) = N(s)/D(s), \tag{3.23}$$

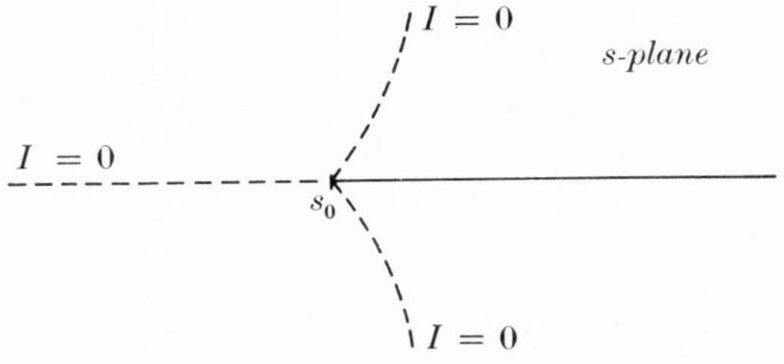

FIG. 3.14. For a P-wave three branches of $I = 0$ meet at s_0 with an angular separation $\frac{2}{3}\pi$. (The scattering length is not zero.)

where, by eqn (2.83) in § 2.6,

$$N(s) = \frac{\Gamma D(P)}{s-P}, \tag{3.24}$$

and $D(s)$ is given by eqn (2.84). It follows, as in § 2.6,† that, as $s \to +\infty$,

$$\operatorname{im} D(s+) \sim s^{\frac{1}{2}},$$

$$\operatorname{re} D(s) = O(1),$$

so

$$\frac{\alpha(\infty)}{\pi} + n_B = \tfrac{1}{2}. \tag{3.25}$$

Theorem 3 (§ 2.4) confirms that this is the isolated solution.

In § 4.2 we shall give the exact solution and the positions of the second sheet poles. Here we show what information can be obtained without going to the explicit solution.

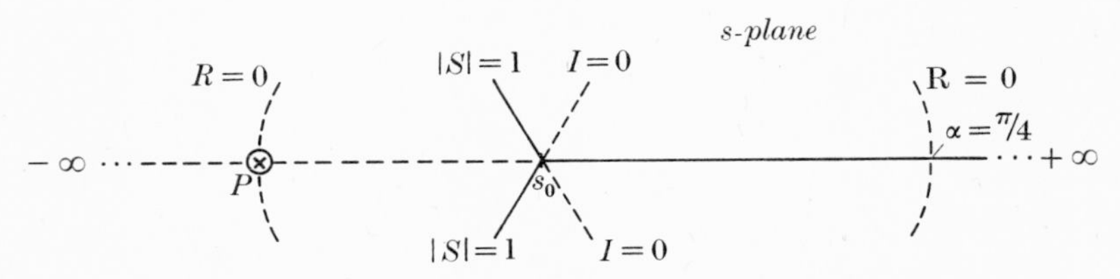

FIG. 3.15. Basic information about the non-resonant P-wave.

We only consider the case in which Γ is not so large that a bound state can occur. Then, by eqn (3.25),

$$\alpha(\infty) = \tfrac{1}{2}\pi.$$

Furthermore

$$D(s)/q(s) \to iC + O(1/q) \tag{3.26}$$

as $s \to +\infty \pm i\epsilon$, where C is a real number. Since the cut $s_0 \leqslant s \leqslant \infty$ is the only singularity of $D(s)$, the limiting behaviour in (3.26) will hold as $|s| \to \infty$. Since

$$S(s) = 1 + 2iq^3 F(s)$$

$$= 1 + 2iq^3 N(s)/D(s), \tag{3.27}$$

and

$$S(s) \to -1 \quad \text{as} \quad s \to +\infty \pm i\epsilon,$$

it follows that

$$S(s) \to -1, \quad \text{as } |s| \to \infty. \tag{3.28}$$

Case (a): no resonance

The phase shift $\alpha(s)$ starts at zero at the threshold s_0 and increases monotonically to reach $\alpha(\infty) = \frac{1}{2}\pi$ (we shall give an argument below suggesting that $\alpha(s)$ must be monotonic). Thus no branch of $|S| = 1$

† The form of eqn (3.24) shows that eqn (2.85) cannot be satisfied.

meets the physical cut except for the branches coming in at s_0. We also know from § 3.5 that three lines $I = 0$ meet at s_0 (Fig. 3.15).

On $-\infty < s < s_0$, $q(s) = i|q(s)|$ and

$$S(s) = 1+2|q|^3F(s).$$

The dispersion relation

$$F(s) = \frac{\Gamma}{s-P}+\frac{1}{\pi}\int\limits_{s_0}^{\infty}\frac{\operatorname{im}F(s')}{s'-s}\,ds'$$

implies that $F(s) > 0$ on $P < s < s_0$. Thus $S(s) > 1$ on $P < s < s_0$ (Fig. 3.16).

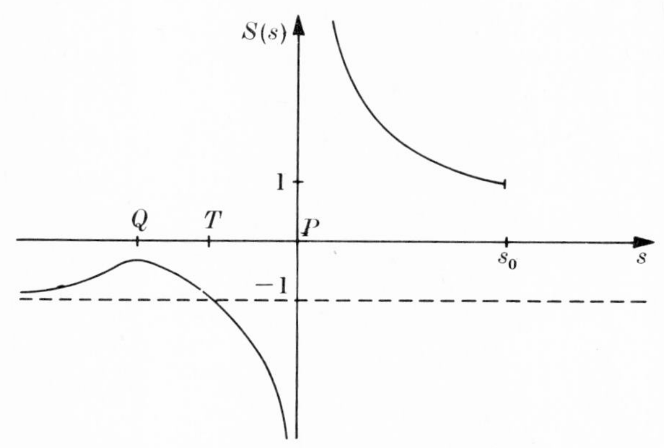

FIG. 3.16. The form of $S(s)$ on $-\infty < s < s_0$ in the non-resonant P-wave.

Using eqns (3.26) and (3.27) it follows that $S(s)$ is regular at infinity in the q-plane. For large $|q|$,

$$S(q) = -1+Q_1/q+Q_2/q^2+\ldots.$$

Only if $Q_1 = 0$ can there be a double point of $|S| = 1$ at infinity, that is, a resonance at infinity ($\operatorname{re}D \to 0$ as $s \to +\infty$). This cannot occur in general, therefore the branch of $|S| = 1$ which comes out of s_0 must meet $-\infty < s < P$ at a finite point. This requires that $S(s) = -1$ at some point T on $-\infty < s < P$. In addition, it requires that $S(s) \to -1$ from above as $s \to -\infty$.

We see from Fig. 3.16 that dS/ds must vanish at a point Q on $-\infty < s < P$, and $Q < T$. Since $S(s)$ is real on $-\infty < s < P$, a branch of $I = 0$ must cross the real axis at Q.

The behaviour of $S(s)$ as $s \to -\infty$ can also be seen as follows. We have

$$q^3\cot\alpha = \frac{\operatorname{re}D(s)}{N(s)} \to \lambda s, \quad \text{as } s \to +\infty,$$

where λ is a constant. Also $\lambda > 0$, since there is no resonance. Then

$$\alpha(s) \to \tfrac{1}{2}\pi - \frac{4\lambda}{q(s)}, \quad \text{as } s \to +\infty,$$

so

$$S(s) \to -1+8\lambda(-\tfrac{1}{4}s)^{-\frac{1}{2}}, \quad \text{as } |s| \to \infty, \tag{3.29}$$

where $(-s)^{\frac{1}{2}}$ is defined in the plane cut along $0 < s < \infty$, so that it is positive on $-\infty < s < 0$. Letting $s \to -\infty$ now gives the result above.

Configuration of the level curves

One possible configuration of the level curves $|S| = 1$, $R = 0$, and $I = 0$ is shown in Fig. 3.17. There is a pair of zeros of S off the real axis,

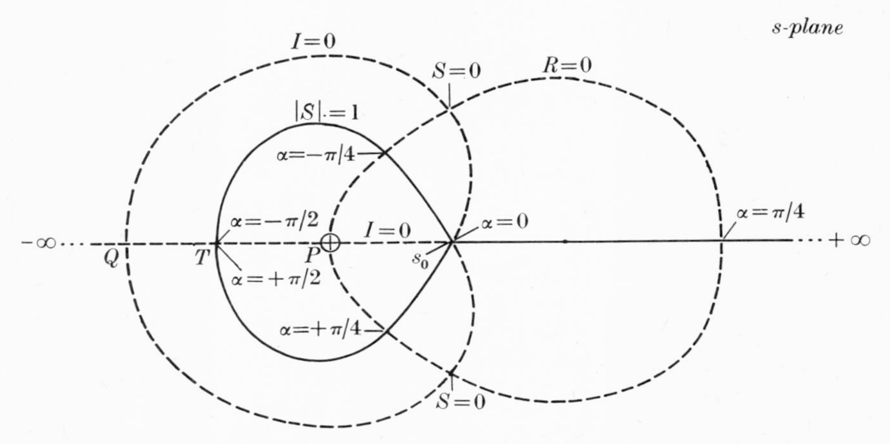

FIG. 3.17. A possible configuration of the level curves of the non-resonant P-wave.

even though there is no resonance. On the branch of $|S| = 1$ near s_0 in $\operatorname{im} s > 0$,

$$\alpha = aq^3 = -a|q|^3 \ (\operatorname{mod} \pi),$$

and since $a > 0$, α decreases as we move along $|S| = 1$ into $\operatorname{im} s > 0$. Starting from $\alpha(s_0) = 0$, the simplest case is $\alpha = -\frac{1}{2}\pi$ at T. Around the closed loop $|S| = 1$, the change in α is $\Delta\alpha = -\pi$, corresponding to one pole inside the loop (eqn (3.19)). Conversely from the fact that there is one pole and no zero inside the loop, we deduce that $\alpha = -\frac{1}{2}\pi$ is the only possibility at T.

In place of Fig. 3.16 we could also have the form in Fig. 3.18. The corresponding level curve configuration is shown in Fig. 3.19. The two zeros of S are now on the real axis $-\infty < s < T$; they are separated by the point Q where a branch of $I = 0$ crosses the axis. We expect that this configuration is given by the small values of Γ. This is verified by the explicit calculation in Chapter 4.

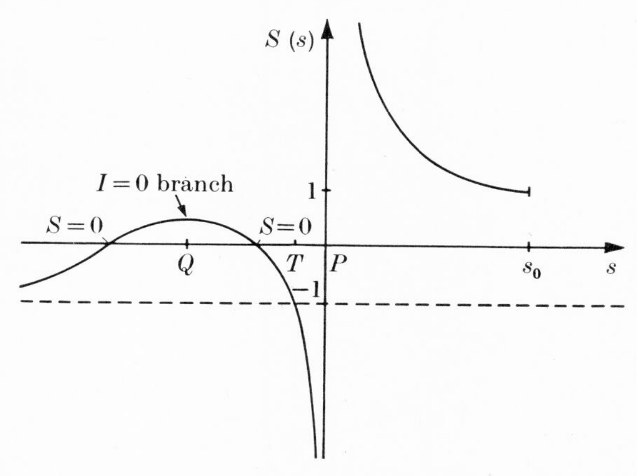

FIG. 3.18. An alternative form to Fig. 3.16 for the non-resonant P-wave.

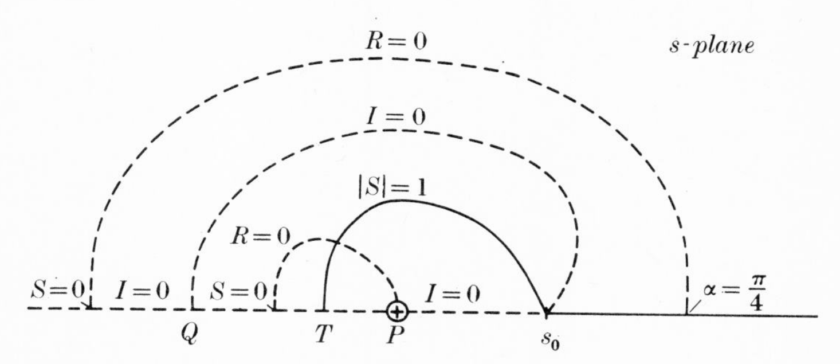

FIG. 3.19. The other possibility for the level curves of the non-resonant P-wave.

The resonant P-wave

In this case α must have a maximum value α_m ($\alpha_m > \frac{1}{2}\pi$) on the physical cut. We shall only consider the case $\frac{1}{2}\pi < \alpha_m < \frac{3}{4}\pi$. A simple form for the level curves is shown in Fig. 3.20. As the coefficient λ in eqn (3.29) is negative when there is a resonance, we conjecture that $S(s) < -1$ on $-\infty < s < P$. Thus there can be no other branch of $|S| = 1$ other than is shown in Fig. 3.20, and no other zero of $S(s)$. Going around the closed loop $|S| = 1$ from s_0, we have $\Delta\alpha = \pi$ corresponding to the zero of S within the loop.

Monotonic nature of α when there is no resonance

The following argument suggests that $\alpha(s)$ must be monotonic increasing on the physical axis if there is no resonance (i.e. if α does not pass through $\frac{1}{2}\pi$). Suppose $\alpha(s)$ has a maximum α_1 ($\alpha_1 < \frac{1}{2}\pi$) at s_1 and a

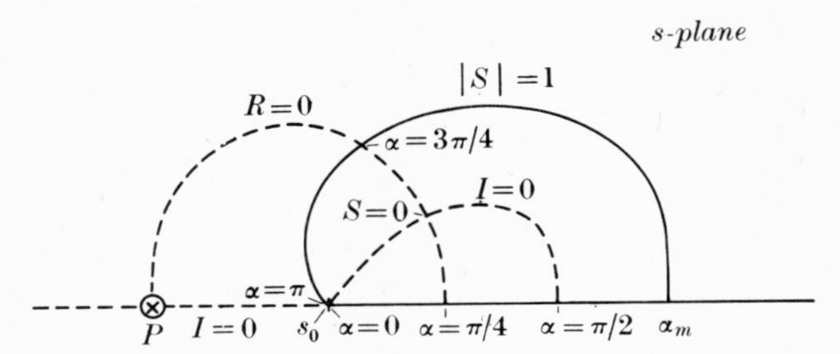

FIG. 3.20. Level curves for the resonant P-wave ($\alpha_m < \frac{3}{4}\pi$).

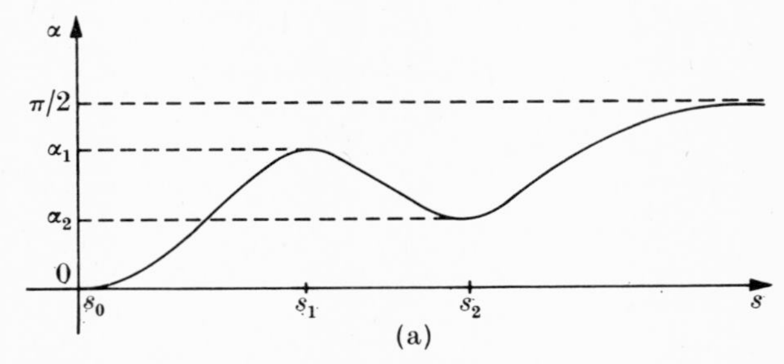

FIG. 3.21 (a). A non-resonant phase having a maximum α_1 at s_1 and a minimum α_2 at s_2.

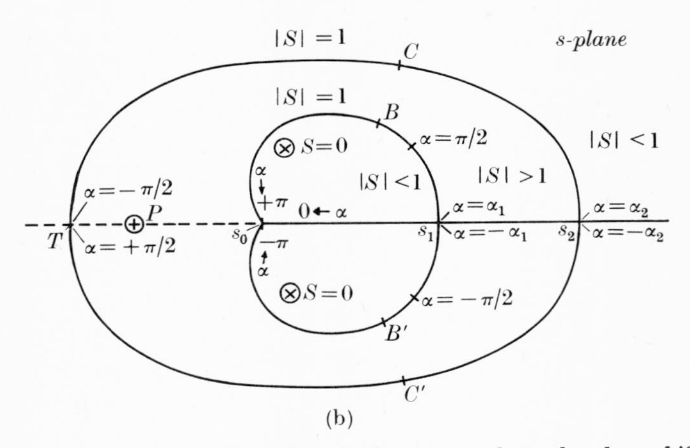

FIG. 3.21 (b). This configuration which corresponds to the phase shift in Fig. 3.21 (a) is impossible.

minimum α_2 ($\alpha_1 > \alpha_2 > 0$) at s_2 on $s_0 < s < \infty$ (Fig. 3.21 (a)). Branches of $|S| = 1$ meet the physical cut at s_0, s_1, and s_2. The simplest configuration is shown in Fig. 3.21 (b). Since $\alpha(s+i\epsilon)$ increases with s on $s_0 < s < s_1$, $|S| < 1$ just above that segment. Thus the closed loop $s_0\, s_1\, Bs_0$ composed of branches of $|S| = 1$ must contain a zero of S. The simplest possibility (i.e. one zero) gives $\Delta\alpha = \pi$ around this loop, and $\alpha \to \pi$ on approaching s_0 along $\arg s = 120°$. We impose the condition eqn (3.7) on $\alpha(s)$, so $\alpha \to -\pi$ on approaching s_0 along $\arg s = 240°$.

The region between the two curved branches of $|S| = 1$ has $|S| > 1$. Also $S = -1$ at some point T on $-\infty < s < P$. Now consider the closed loop $\mathscr{C} \equiv s_0 B s_1 s_2 C T C' s_2 s_1 B' s_0$ formed of branches of $|S| = 1$. Since there is one pole and no zero inside, on going around the loop we should have $\Delta\alpha = -\pi$.

On going from s_1 to s_2 α decreases, therefore α decreases on moving along $s_2 CT$. We take α to be $-\frac{1}{2}\pi$ on approaching T from above (our argument would be even stronger if the value were $-\frac{3}{2}\pi$). Starting at s_0 along $\arg s = 120°$ and ending at s_0 along $\arg s = 240°$ we see that around the loop $\mathscr{C}$, $\Delta\alpha = -3\pi$. The contradiction shows that the configuration in Fig. 3.21 (b) is impossible.

4

POLE REGIONS FOR ATTRACTIVE *P*-WAVES: EQUAL MASS HEAVY PARTICLES

4.1. Introduction

IN this chapter we give expressions which determine the positions of the resonance poles of a P-wave, or else tell us the region in which these poles can lie. We treat the equal mass heavy particle case; as we pointed out in Chapter 1, in this (e.m.h.) case, im $f_l(s)$ is specified on the left-hand cut $-\infty < s < s_1$. Here we use the kinematical relation $q^2 = \frac{1}{4}(s-s_0)$, and we should emphasize that this is not a non-relativistic approximation.

We shall principally consider elastic scattering, although in the final section the modifications required in the inelastic case where $\eta(s)$ is specified are described. We consider P-waves having an attractive interaction because this is the simplest type in which resonances occur naturally.

Our objective is to trace the motion of the second sheet poles as the strength of the interaction is increased, and in general we shall not allow the strength to exceed the value at which the first bound state appears. The relation between the resonance poles and the resonance position (where $\alpha = \frac{1}{2}\pi$) will also be discussed.

4.2. *P*-wave (e.m.h. case): single attractive pole

This problem has been examined by various authors, [22], [23], and pp. 149–51 of [21], but we give it here because it has some features characteristic of the general case. Also it is a simple way of introducing the notation (see also § 3.6).

The kinematical relation is

$$q^2 = \tfrac{1}{4}(s-s_0), \tag{4.1}$$

where $s_0 = 4m^2$ is the physical threshold. We assume the scattering is elastic, so the reduced P-wave amplitude $F(s) = f(s)/q^2$ has the form

$$F(s+i\epsilon) = \frac{\exp(i\alpha(s))\sin\alpha(s)}{q^3}, \quad \text{for } s_0 \leqslant s \leqslant \infty,$$

on the physical cut. The isolated solution is used and the N/D equations are given in eqns (2.83) and (2.84) of § 2.6 above.

There is a single driving pole

$$\Gamma/(s-P), \quad \text{with } -\infty < P < s_0,$$

in $F(s)$. The pole is attractive if $\Gamma > 0$. Equation (2.83) gives

$$N(s) = \frac{\Gamma D_p}{s-P}, \tag{4.2}$$

where $D_p \equiv D(P)$. Integrating eqn (2.84) gives

$$D(s) = D_0 - \frac{s-s_0}{s-P}\frac{1}{8}p\Gamma D_p - iq^3\frac{\Gamma D_p}{s-P} \tag{4.3}$$

for any s on the physical sheet. Here $q(s) = \frac{1}{2}(s-s_0)^{\frac{1}{2}}$ is defined as in Fig. 3.1 (§ 3.1). On $-\infty < s < s_0$, $q(s) = \frac{1}{2}i|s_0-s|^{\frac{1}{2}}$. We have written $p = +(s_0-P)^{\frac{1}{2}}$, and the notation is further simplified by defining

$$\gamma = \tfrac{1}{8}\Gamma p.$$

Equation (4.3) gives

$$D_p = D_0/(1-\tfrac{1}{2}\gamma). \tag{4.4}$$

In order to keep D_p finite as $\gamma \to 2$ we should have to put $D_0 \equiv D(s_0) = 0$. This means that $D(s)$ vanishes at s_0 for $\gamma = 2$; i.e. there is a resonance or bound state at the threshold. We are mainly interested in values of Γ below this critical value, i.e. in the range $0 < \gamma < 2$.

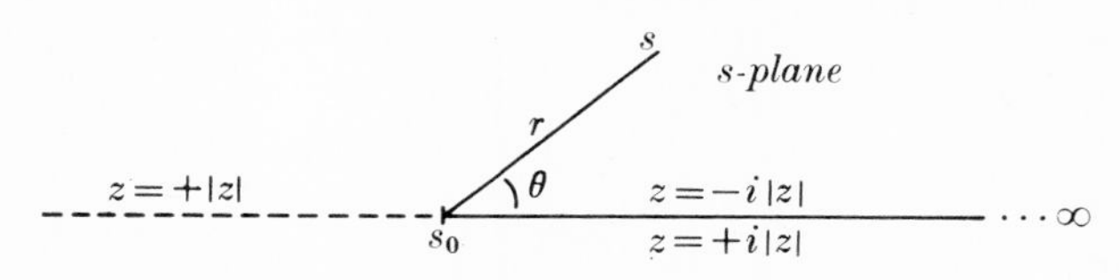

FIG. 4.1. The definition of $z = (s_0-s)^{\frac{1}{2}}$.

It is convenient to replace the variable s by

$$z = (s_0-s)^{\frac{1}{2}}$$

defined in the s-plane cut along $s_0 < s < \infty$, as in Fig. 4.1. On the physical sheet

$$2q = iz, \tag{4.5}$$

and, at $s = s_0+r\exp(i\theta)$,

$$\left.\begin{aligned} z &= -ir^{\frac{1}{2}}\exp(\tfrac{1}{2}i\theta) \\ \text{so} \qquad \operatorname{re} z &= r^{\frac{1}{2}}\sin(\tfrac{1}{2}\theta) \end{aligned}\right\}. \tag{4.6}$$

Thus the physical sheet is defined by

$$\operatorname{re} z \geqslant 0. \tag{4.7}$$

The physical cut $s_0 \leqslant s \leqslant \infty$ is given by the line $\operatorname{re} z = 0$; the upper side of this cut maps on to the line $\operatorname{re} z = 0$, $\operatorname{im} z \leqslant 0$, and the lower side maps on to $\operatorname{re} z = 0$, $\operatorname{im} z \geqslant 0$. The transformation $z \to -z$ relates the point s on the physical sheet $(0 < \theta < 2\pi)$ to the same value of s on the second sheet $(2\pi < \theta < 4\pi)$. The whole z-plane describes the physical sheet and the second sheet of the s-plane.

The S-function is

$$S(s) = \frac{D(s)+2iq^3N(s)}{D(s)}. \tag{4.8}$$

Equations (4.2), (4.3), and (4.4) give, for $\operatorname{re} z \geqslant 0$,

$$S(s) = \frac{p^2(1-\frac{1}{2}\gamma)-z^2(1-\frac{3}{2}\gamma)+\gamma z^3/p}{p^2(1-\frac{1}{2}\gamma)-z^2(1-\frac{3}{2}\gamma)-\gamma z^3/p}. \tag{4.9}$$

The meromorphic form of $S(z)$

By eqn (3.11) the continuation $S^{\mathrm{II}}(s)$ of $S(s)$ on to the second sheet obeys $S^{\mathrm{II}}(s) = (S(s))^{-1}$. Thus on writing S as a function of z it must obey

$$S(-z) = (S(z))^{-1}. \tag{4.10}$$

This relation, which is quite general, is obviously satisfied by eqn (4.9), which therefore gives $S(s)$ on both sheets.

The driving pole is at $s = P$, i.e. $z = p$. Therefore $(z-p)$ must be a factor of the denominator in eqn (4.9). The remaining zeros of the denominator give the remaining poles of $S(s)$. If they obey $\operatorname{re} z \leqslant 0$ they are the second sheet poles of $S(s)$, i.e. the second sheet poles of $F(s)$.

These remaining zeros obey

$$\gamma z^2+pz(1-\tfrac{1}{2}\gamma)+p^2(1-\tfrac{1}{2}\gamma) = 0. \tag{4.11}$$

We shall call this the *pole equation.*

Let the zeros be z_1 and z_2. Then by eqn (4.10) we have

$$S(z) = -\frac{z+p}{z-p}\cdot\frac{(z+z_1)(z+z_2)}{(z-z_1)(z-z_2)}.$$

Another useful form comes from eqn (4.8) and the explicit forms of $D(s)$ and $N(s)$ given in eqns (4.2) and (4.3). Writing D as a function of z and using eqn (4.5) we get

$$D(z)+2iq^3N(z) = D(-z).$$

Therefore

$$S(z) = D(-z)/D(z). \tag{4.12}$$

The second sheet poles of $F(s)$ are given by the zeros of

$$D(z) = 0, \tag{4.13}$$

which obey $\operatorname{re} z < 0$.

Bound state and resonance

For $\gamma \to 2$ the roots of the pole equation (4.11) tend to zero. Then there is a pole (or resonance) at the threshold s_0. The roots of eqn (4.11) are

$$z = -\frac{p}{2\gamma}[(1-\tfrac{1}{2}\gamma)\pm\{(1-\tfrac{1}{2}\gamma)(1-\tfrac{9}{2}\gamma)\}^{\frac{1}{2}}]. \tag{4.14}$$

For $\gamma > 2$ both roots are real, one being positive and the other negative. The positive root gives a pole of $S(s)$ or $F(s)$ on the physical sheet; this is the bound state pole. The negative root gives the second sheet pole of $F(s)$, which lies between s_0 and the bound state pole.

We define a resonance by requiring that $\alpha(s)$ should pass upwards through $\frac{1}{2}\pi$ on the physical cut at a finite energy, and the resonance position is given by s_R where $\alpha(s_R) = \frac{1}{2}\pi$. Now

$$q^3 \cot\alpha = \operatorname{re} D/N, \quad \text{for } s_0 \leqslant s \leqslant \infty.$$

By eqns (4.2) and (4.3),

$$q^3 \cot\alpha = \frac{1}{\Gamma}\{(s-s_0)(1-\tfrac{3}{2}\gamma)+p^2(1-\tfrac{1}{2}\gamma)\}. \tag{4.15}$$

The scattering length is

$$a = \frac{8}{p^3}\frac{\gamma}{1-\frac{1}{2}\gamma}, \tag{4.16}$$

and it is positive for $0 < \gamma < 2$.

The coefficient of the term $(s-s_0)$ on the right of eqn (4.15) shows that there is no resonance for $0 < \gamma < \frac{2}{3}$. For $\frac{2}{3} < \gamma < 2$ there is a resonance, and

$$s_R - s_0 = p^2\frac{1-\frac{1}{2}\gamma}{\frac{3}{2}\gamma-1}. \tag{4.17}$$

As $\gamma \to \frac{2}{3}$, $s_R \to +\infty$, and as $\gamma \to 2$, $s_R \to s_0$. For $\gamma = 1$, $s_R - s_0 = p^2 = s_0 - P$.

The motion of the poles

We shall examine the motion of the roots of eqn (4.11) as γ increases from 0 to 2. Equation (4.14) shows that both roots lie in $\operatorname{re} z < 0$ for $0 \leqslant \gamma < 2$, therefore both the poles are on the second sheet.

(*a*) $0 \leqslant \gamma \leqslant \frac{2}{9}$

Equation (4.14) shows that both roots z_1 and z_2 are real (and negative) in this range of γ. For small γ,

$$\left.\begin{aligned} z_1 &\to -p/\gamma \\ z_2 &\to -p \end{aligned}\right\}, \quad \text{as } \gamma \to 0.$$

Equation (4.11) can only have the root $z = -p$ for $\gamma = 0$. For $0 < \gamma \leqslant \frac{2}{9}$ the two poles lie on $-\infty < s < P$. For $\gamma = \frac{2}{9}$, $z_1 = z_2 = -2p$ (Fig. 4.2).

(*b*) $\frac{2}{9} < \gamma < 2$

In this range the roots of eqn (4.11) are a complex conjugate pair. We call them $\tilde{z}$ and $\tilde{z}^*$, and write

$$\left.\begin{aligned} \tilde{z} &= |\tilde{z}|\exp(i\tilde{\psi}) \\ \text{where} \qquad \tilde{\psi} &= \tfrac{1}{2}(\theta-\pi) \end{aligned}\right\}, \tag{4.18}$$

by eqn (4.6). From eqn (4.11)

$$\left.\begin{aligned} |\tilde{z}|^2 &= p^2\left(\frac{1}{\gamma}-\frac{1}{2}\right) \\ 2\operatorname{re}\tilde{z} &= -p\left(\frac{1}{\gamma}-\frac{1}{2}\right) \end{aligned}\right\}, \tag{4.19}$$

for $\frac{2}{9} < \gamma < 2$. Eliminating γ gives the locus

$$|\tilde{z}| = -2p\cos\tilde{\psi}. \tag{4.20}$$

This is a circle of radius p whose centre is $(-p)$. It is entirely in $\operatorname{re} z \leqslant 0$, and touches the line $\operatorname{re} z = 0$ at $z = 0$. As γ increases from $\frac{2}{9}$ to 2, $\tilde{z}$ moves monotonically around the circle from $-2p$ to 0.

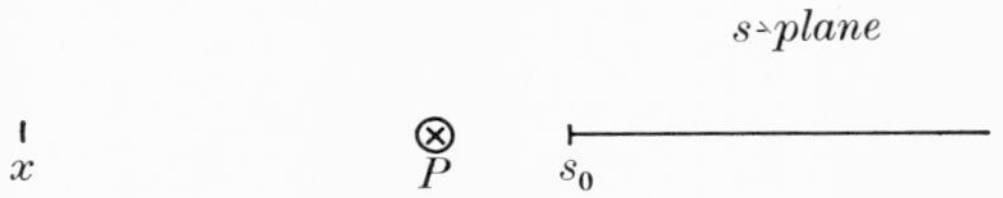

FIG. 4.2. The two zeros of $S(s)$ coincide at x for $\gamma = \frac{2}{9}$, where $s_0-x = 4(s_0-P)$. For $\gamma > \frac{2}{9}$ they move off into the upper and lower half planes.

If $\tilde{s}$ is the position of the second sheet pole, eqn (4.20) gives the locus

$$|\tilde{s}-s_0| = 4p^2\sin^2(\tfrac{1}{2}\theta). \tag{4.21}$$

This is a cardioid curve as shown in Fig. 4.3. As γ increases from $\frac{2}{9}$ to 2, $\tilde{s}$ moves around the cardioid from s_0-4p^2 to s_0. The curve is tangential to the physical cut at s_0.

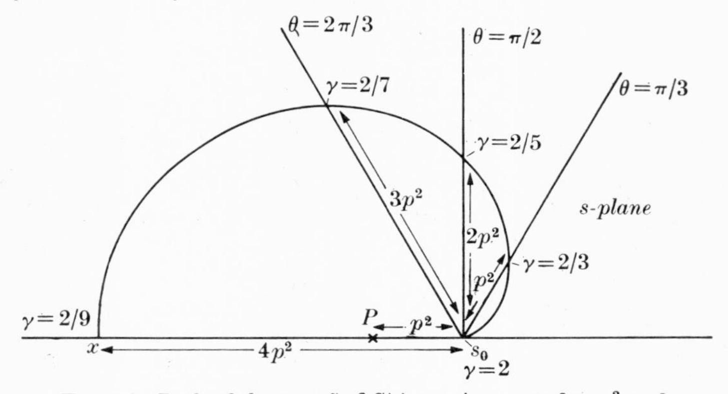

FIG. 4.3. Path of the zero $\tilde{s}$ of $S(s)$ as γ increases from $\frac{2}{9}$ to 2.

From eqn (4.19) we find that the angular position of the second sheet pole in the s-plane is given by

$$\cos\theta = \frac{5}{4} - \frac{1}{2\gamma}, \quad \text{for } \tfrac{2}{9} \leqslant \gamma \leqslant 2. \tag{4.22}$$

For $\gamma = \frac{2}{3}$, $\cos\theta = \frac{1}{2}$, so $\tilde{s}$ lies on the line $\theta = \frac{1}{3}\pi$ in Fig. 4.3. We saw that for $\frac{2}{9} < \gamma < \frac{2}{3}$ there is no resonance, but nevertheless $F(s)$ has a pair of second sheet poles lying away from the real axis. For $\frac{2}{3} < \gamma < 2$ there is a resonance, but it is clear that the poles $\tilde{s}$ (and $\tilde{s}^*$) are only near the resonance position for $1 \lesssim \gamma < 2$. For $\gamma = 1$, eqns (4.17) and (4.19) give

$$s_R - s_0 = p^2, \quad |\tilde{s} - s_0| = \tfrac{1}{2}p^2, \quad \text{and} \quad \theta \simeq 41^\circ,$$

so $\tilde{s}$ and s_R are less than p^2 apart when $\gamma = 1$.

By eqns (4.21) and (4.22),

$$\operatorname{re}\tilde{s} - s_0 = p^2\left(\frac{1}{\gamma} - \frac{1}{2}\right)\left(\frac{5}{4} - \frac{1}{2\gamma}\right).$$

Therefore

$$\max(\operatorname{re}\tilde{s} - s_0) = \tfrac{1}{2}p^2 \tag{4.23}$$

and the maximum occurs for $\gamma = \frac{2}{3}$, $\theta = \frac{1}{3}\pi$.

Similarly using eqn (4.17)

$$s_R - \operatorname{re}\tilde{s} = \frac{(2-\gamma)^2}{3\gamma - 2}\,\frac{7\gamma - 2}{8\gamma^2}\,p^2.$$

For γ just above $\frac{2}{3}$,

$$s_R - \operatorname{re}\tilde{s} \simeq \frac{\frac{4}{9}p^2}{\gamma - \frac{2}{3}} \quad (\gamma \gtrsim \tfrac{2}{3}), \tag{4.24}$$

and for γ just under 2,

$$s_R - \operatorname{re}\tilde{s} \simeq \tfrac{3}{32}p^2(2-\gamma)^2 \quad (\gamma \lesssim 2). \tag{4.25}$$

Thus when γ is near to 2, the pole position and s_R are close together since

$$\frac{\operatorname{im}\tilde{s}}{|\tilde{s} - s_0|} \simeq \tfrac{1}{2}(2-\gamma)^{\frac{1}{2}} \quad (\gamma \lesssim 2). \tag{4.26}$$

In Fig. 4.4 we illustrate the form of the resonance in two cases by plotting $\sin^2\alpha$ for $s_R - s_0 = \frac{2}{3}p^2$, i.e. $\gamma = \frac{10}{9}$, $\tilde{s} - s_0 = \frac{2}{25}p^2(4+3i)$, and for $s_R - s_0 = \frac{1}{9}p^2$, i.e. $\gamma = \frac{5}{3}$, $\tilde{s} - s_0 = \frac{1}{200}p^2(19 + i\sqrt{39})$.

4.3. The unitary limit

It has been conjectured by Donnachie and Hamilton [24] that the left-hand cut term in the dispersion relation for a reduced p.w.a. $F(s) = f(s)/q^{2l}$ (with orbital angular momentum $l \geqslant 1$) obeys an

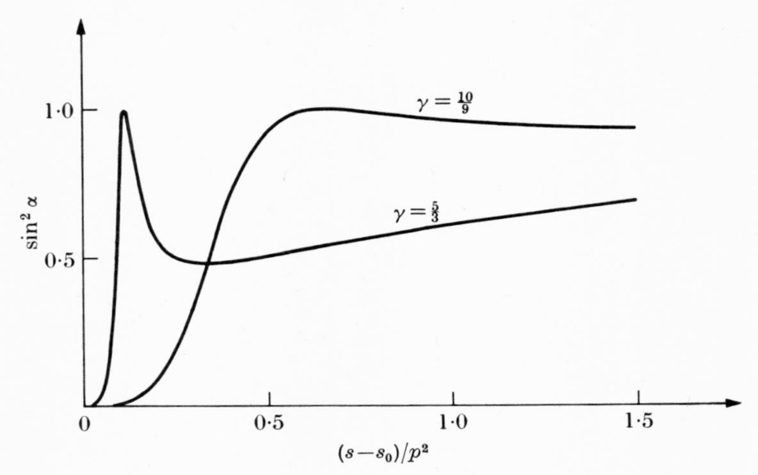

FIG. 4.4. Plot of $\sin^2\alpha(s)$ for $\gamma = \frac{10}{9}$, and for $\gamma = \frac{5}{3}$.

inequality at the resonance position. Writing the dispersion relation in the form

$$\text{re}\,F(s) = \bar{F}(s) + \frac{1}{\pi}P\int\limits_{s_0}^{\infty}\frac{\text{im}\,F(s')}{s'-s}\,ds', \quad \text{for } s_0 \leqslant s \leqslant \infty,$$

the conjecture is that

$$q_R^{2l+1}\bar{F}(s_R) < \tfrac{1}{2}, \tag{4.27}$$

where $q_R = q(s_R)$. The argument is based on the idea that since by unitarity $q^{2l+1}\,\text{re}\,F(s) \leqslant \frac{1}{2}$, the saturation of unitarity at the resonance requires a value of $q_R^{2l+1}\bar{F}(s_R)$ which, though not small, does not violate eqn (4.27) (see pp. 251–9 of [24]). The rule is obeyed by those πN resonances for which $\bar{F}(s)$ is accurately known (i.e. N_{33}^*, N_{13}^*, N_{15}^*, N_{37}^*).

In the case of the single attractive pole P-wave which has been discussed in the preceding section this rule is not obeyed in general. We have

$$\bar{F}(s) = \Gamma/(s-P),$$

and using eqns (4.1) and (4.17) we get

$$q_R^3\,\bar{F}(s_R) = (1-\tfrac{1}{2}\gamma)^{\frac{3}{2}}(\tfrac{3}{2}\gamma-1)^{-\frac{1}{2}}. \tag{4.28}$$

The right-hand side of eqn (4.28) is less than $\frac{1}{2}$ for $2 > \gamma > 1$, but it exceeds $\frac{1}{2}$ if $\gamma < 1$, and it becomes infinite as $\gamma \to \frac{2}{3}$. It is useful to remember that $s_R - s_0 = p^2$ for $\gamma = 1$, and also that $\theta = 41°$; so for $\gamma = 1$ the resonance is no longer narrow.

Since s_R is not always closely related to the pole position $\tilde{s}$, it may seem useful to modify the unitary limit conjecture to apply it to the position of the second sheet pole. For $l = 1$ we write

$$L(s) \equiv q^3(s)\bar{F}(s),$$

and we *conjecture* that for a resonance

$$|L(\tilde{s})| < \tfrac{1}{2}. \tag{4.29}$$

It is easy to test this inequality. From eqns (4.19), (4.20), (4.21), and (4.22)

$$|\tilde{s}-P| = p^2(2/\gamma)^{\frac{1}{2}}, \quad \text{for } \tfrac{2}{9} < \gamma < 2,$$

and we get

$$|L(\tilde{s})| = \tfrac{1}{4}(2-\gamma)^{\frac{3}{2}}. \tag{4.30}$$

Therefore $|L(\tilde{s})| < \frac{1}{2}$ for $0{\cdot}41 < \gamma < 2$, so eqn (4.29) is certainly valid in the whole resonance region $\frac{2}{3} < \gamma < 2$.

If the conjecture in eqn (4.29) were true for the general left-hand discontinuity we would have another method of some value for locating the position of resonance poles. If in addition we knew that the resonance was not wide, the conjecture would give an upper limit on the resonance position s_R.

4.4. *P*-wave (e.m.h. case): truncated left-hand cut

We now extend the analysis of § 4.2 to deal with a left-hand cut discontinuity

$$\rho(s) \equiv \operatorname{im} F(s+i\epsilon)$$

such that $\rho(s)$ vanishes outside a segment $s_2 \leqslant s \leqslant s_1$, where $s_1 < s_0$. Again the isolated solution is used and the N/D equations are given in eqns (2.83) and (2.84). We put $D_0 \equiv D(s_0) = 1$. First we discuss the general properties of $D(z)$.

Analytic continuation of $D(z)$

The function $N(s)$ is regular in the s-plane cut along $s_2 \leqslant s \leqslant s_1$, so expressed as a function of z, $N(z)$ is regular in the half-plane $\operatorname{re} z \geqslant 0$ cut along $p_1 \leqslant z \leqslant p_2$, where

$$p_1 = (s_0-s_1)^{\frac{1}{2}}, \quad p_2 = (s_0-s_2)^{\frac{1}{2}}.$$

The function $D(z)$ is regular in the whole region $\operatorname{re} z \geqslant 0$.

Let $H(z)$ be defined in $\operatorname{re} z \leqslant 0$ by

$$H(-z) = D(z)+2iq^3(z)N(z), \quad \text{for } \operatorname{re} z \geqslant 0. \tag{4.31}$$

Thus $H(z)$ is regular in the region $\operatorname{re} z < 0$ cut along $-p_2 \leqslant z \leqslant -p_1$. On $\operatorname{re} z = 0$,

$$\operatorname{im} N(z) = 0, \quad \operatorname{im} D(z) = -q^3(z)N(z);$$

therefore

$$\left.\begin{aligned} \operatorname{re} H(-z) &= \operatorname{re} D(z) \\ \operatorname{im} H(-z) &= -\operatorname{im} D(z) \end{aligned}\right\} \quad \text{on } \operatorname{re} z = 0.$$

By eqns (2.84) and (4.6), $D(s^*) = D^*(s)$ and $z(s^*) = z^*(s)$. Therefore

$$D(z^*) = D^*(z).$$

It follows that on $\operatorname{re} z = 0$,

$$H(-z) = D^*(z) = D(-z).$$

Therefore $H(z)$ is the analytic continuation of $D(z)$ into the half-plane $\operatorname{re} z \leqslant 0$. Thus $D(z)$ is regular in the z-plane cut along $-p_2 \leqslant z \leqslant -p_1$.

By eqns (4.5) and (4.31), for $\operatorname{re} z \geqslant 0$,

$$D(-z) = D(z) + \tfrac{1}{4}z^3 N(z). \tag{4.32}$$

It follows that $N(z)$ is an even function of z on $\operatorname{re} z = 0$. Therefore $N(z)$ is an even function of z and is regular in the z-plane cut along $-p_2 \leqslant z \leqslant -p_1$, $p_1 \leqslant z \leqslant p_2$. As a consequence, eqn (4.32) is valid in the whole z-plane.

Finally we notice that near $z = 0$

$$D(z) = 1 + c_2 z^2 + c_3 z^3 + \ldots, \tag{4.33}$$

where $c_2, c_3, \ldots$ are real constants. There is no term linear in z on the right of eqn (4.33) because on $\operatorname{re} z = 0$, $\operatorname{im} D(z) = O((\operatorname{im} z)^3)$ near $z = 0$.

The dispersion relation for $D(z)$

Because the left-hand cut is truncated it follows that

$$|N(s)| = O(|s|^{-1}), \quad \text{as } |s| \to \infty,$$

in the physical sheet. Therefore in the z-plane

$$|N(z)| = O(|z|^{-2}), \quad \text{as } |z| \to \infty.$$

The unitarity condition

$$\operatorname{im} D(s+) = -q^3 N(s), \quad \text{for } s_0 \leqslant s \leqslant \infty,$$

gives

$$\operatorname{im} D(s+) = O(s^{\frac{1}{2}}), \quad \text{as } s \to +\infty.$$

By eqn (2.84) and eqns (I.1) and (I.3) of Appendix I it follows that

$$\operatorname{re} D(s) = O(1), \quad \text{as } s \to +\infty,$$

and

$$|D(s)| = O(|s|^{\frac{1}{2}}), \quad \text{as } |s| \to \infty.$$

Now using eqn (4.32) we find that

$$|D(z)| = O(|z|), \quad \text{as } |z| \to \infty,$$

in the whole z-plane.

The function

$$\frac{D(z)-1}{z^2}$$

is regular at $z = 0$ (compare with eqn (4.33)) and

$$\left|\frac{D(z)-1}{z^2}\right| = O(|z|^{-1}), \quad \text{as } |z| \to \infty.$$

Therefore we have the dispersion relation

$$\frac{D(z)-1}{z^2} = \frac{1}{\pi}\int\limits_{-p_2}^{-p_1} \frac{\operatorname{im} D(z'+)}{z'^2(z'-z)}\,dz'.$$

By eqn (4.32), for $-p_2 \leqslant z \leqslant -p_1$,

$$\begin{aligned}\operatorname{im} D(z+) &= -\tfrac{1}{4}z^3 \operatorname{im} N(z+)\\ &= -\tfrac{1}{4}z^3 \operatorname{im} N(-z-).\end{aligned}$$

Using eqns (2.83) and (4.6) gives

$$\operatorname{im} D(z+) = -\tfrac{1}{4}z^3\rho(s_0-z^2)D(-z), \quad \text{for } -p_2 \leqslant z \leqslant -p_1. \tag{4.34}$$

The dispersion relation can now be written

$$D(z) = 1-\frac{z^2}{4\pi}\int\limits_{-p_2}^{-p_1} \frac{z'\rho(s_0-z'^2)D(-z')}{z'-z}\,dz',$$

or

$$D(z) = 1-\frac{z^2}{4\pi}\int\limits_{p_1}^{p_2} \frac{z'\rho(s_0-z'^2)D(z')}{z'+z}\,dz'. \tag{4.35}$$

For $p_1 \leqslant z \leqslant p_2$ this is an integral equation. Its solution gives $D(z)$ in that interval. Having determined $D(z)$ for $p_1 \leqslant z \leqslant p_2$, eqn (4.35) gives $D(z)$ in the whole z-plane.

Using eqn (4.32) gives us eqn (4.12). The zeros of $D(z)$ in $\operatorname{re} z < 0$ give us the second sheet poles of $F(s)$.

4.5. Attractive left-hand cut discontinuity

The left-hand cut is from now on restricted to be attractive, i.e. we have $\rho(s) \leqslant 0$ on $s_2 \leqslant s \leqslant s_1$ (and $\rho(s) = 0$ elsewhere). In order to vary the strength of the interaction while keeping its 'profile' unaltered we write

$$\rho(s) = -\lambda\bar{\rho}(z), \quad \text{for } s_2 \leqslant s \leqslant s_1,$$

where $\bar{\rho}(z) \geqslant 0$, and λ can vary in $0 < \lambda < \infty$. Equation (4.35) becomes

$$D(z,\lambda) = 1+\frac{\lambda}{4\pi}z^2\int\limits_{p_1}^{p_2} \frac{z'\bar{\rho}(z')D(z',\lambda)}{z'+z}\,dz'. \tag{4.36}$$

Clearly $D(z,0) = 1$. In Appendix III we show that a positive number λ_1 exists, such that the solution $D(z,\lambda)$ of eqn (4.36) is a monotonically increasing function of λ in $0 \leqslant \lambda < \lambda_1$, for fixed z in $p_1 \leqslant z \leqslant p_2$. Furthermore

$$D(z,\lambda) \to +\infty \quad \text{as } \lambda \to \lambda_1, \quad \text{for } p_1 \leqslant z \leqslant p_2.$$

Clearly $D(z,\lambda) > 1$ for $0 < \lambda < \lambda_1$ and $p_1 \leqslant z \leqslant p_2$. The value $\lambda = \lambda_1$

corresponds to a bound state (or a resonance) occurring at the threshold s_0. Our problem is to locate the poles of the S-function (i.e. the zeros of $D(z,\lambda)$) for $0<\lambda<\lambda_1$.

Consider the real positive functions $a_i(z)$:

$$a_i(z) = \frac{\lambda}{4\pi}\int_{p_1}^{p_2} \frac{(z')^i\bar{\rho}(z')D(z',\lambda)}{|z'+z|^2}\,\mathrm{d}z' \quad (i = 1,2,3,\ldots). \tag{4.37}$$

These functions are defined except at the points of $-p_2 \leqslant z \leqslant -p_1$ for which $\bar{\rho}(-z) \neq 0$ (or $\bar{\rho}(-z')$ does not approach zero sufficiently fast as $z' \to z$). We shall denote by E the set of points on $-p_2 \leqslant z \leqslant -p_1$ for which the $a_i(z)$ are not defined.

Since

$$\frac{a_{i+j}(z)}{a_i(z)} = \int_{p_1}^{p_2} (z')^j \frac{(z')^i\bar{\rho}(z')D(z',\lambda)}{|z'+z|^2}\,\mathrm{d}z' \Bigg/ \int_{p_1}^{p_2} \frac{(z')^i\bar{\rho}(z')D(z',\lambda)}{|z'+z|^2}\,\mathrm{d}z',$$

we have

$$p_1^j \leqslant \frac{a_{i+j}(z)}{a_i(z)} \leqslant p_2^j \tag{4.38}$$

for all z, all positive integers i, j, and $0<\lambda<\lambda_1$.

In this notation, eqn (4.36) becomes

$$D(z,\lambda) = 1+z|z|^2a_1(z)+z^2a_2(z). \tag{4.39}$$

Writing $z = |z|\exp(i\psi)$ where $\psi = \frac{1}{2}(\theta-\pi)$ (compare with eqn (4.18)) we have

$$\operatorname{im} D(z,\lambda) = |z|^3a_1(z)\sin\psi + |z|^2a_2(z)\sin 2\psi. \tag{4.40}$$

The curve $\operatorname{im} D(z,\lambda) = 0$

The locus $\operatorname{im} D(z,\lambda) = 0$ has two branches:

(*a*) $\sin\psi = 0$, except the set E on $-p_2 \leqslant z \leqslant -p_1$, i.e. the line $\operatorname{im} z = 0$ except the set E on $-p_2 \leqslant z \leqslant -p_1$. This gives the real axis $-\infty \leqslant s \leqslant s_0$ on the physical sheet and the same line on the second sheet except the portion of $s_2 \leqslant s \leqslant s_1$, which corresponds to the set E.

(*b*) the curve

$$|z| = -2\frac{a_2(z)}{a_1(z)}\cos\psi. \tag{4.41}$$

By eqn (4.38), $\quad p_1 \leqslant a_2(z)/a_1(z) \leqslant p_2,$

therefore for all λ in $0<\lambda<\lambda_1$, the branch given by eqn (4.41) lies in the region bounded by the curves

$$|z| = -2p_1\cos\psi, \qquad |z| = -2p_2\cos\psi. \tag{4.42}$$

This region lies in $\operatorname{re} z \leqslant 0$ and it is illustrated in Fig. 4.5. For any λ in $0<\lambda<\lambda_1$, the branch of $\operatorname{im} D(z,\lambda) = 0$ given by eqn (4.41) passes through $z = 0$ (as $\psi \to \frac{1}{2}\pi$). Also each line $\psi =$ constant $(\frac{1}{2}\pi < \psi < \frac{3}{2}\pi)$

must cross this branch at a point (other than $z = 0$) in the region defined by eqns (4.42) (i.e. the shaded region in Fig. 4.5).

The zeros of $D(z)$ in $\text{re}\, z \leqslant 0$ are the second sheet poles of $F(s)$. When these are complex (and $0 < \lambda < \lambda_1$) they must obey eqn (4.41). In the s-plane these poles must lie in the region bounded by the cardioids

$$\left.\begin{aligned} |s-s_0| &= 4|s_0-s_1|\sin^2(\tfrac{1}{2}\theta) \\ \text{and} \qquad |s-s_0| &= 4|s_0-s_2|\sin^2(\tfrac{1}{2}\theta) \end{aligned}\right\}. \tag{4.43}$$

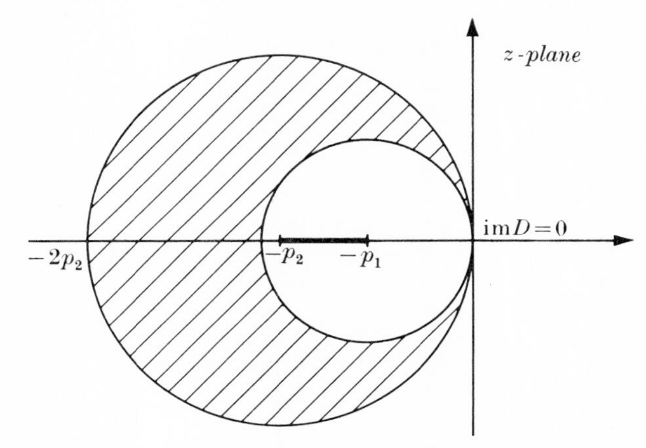

FIG. 4.5. The second branch of im $D = 0$ has to lie in the shaded region.

From the maximum possible value of $|s-s_0|\cos\theta$, we find that the position $\tilde{s}$ of the second sheet pole obeys

$$\text{re}\,\tilde{s} \leqslant s_0+\tfrac{1}{2}p_2^2 = s_0+\tfrac{1}{2}(s_0-s_2). \tag{4.44}$$

On comparing eqns (4.43) and (4.44) with eqns (4.21) and (4.23) it is clear that eqns (4.43) and (4.44) are the optimum which can be obtained without further specifying the discontinuity $\rho(s)$ on $s_2 \leqslant s \leqslant s_1$.

4.6. At most one pair of complex conjugate poles

Again we assume $\rho(s) \leqslant 0$ on $s_2 \leqslant s \leqslant s_1$, and $\rho(s) = 0$ elsewhere. We shall show that if $0 < \lambda < \lambda_1$, then $F(s)$ has at most one pair of complex conjugate second sheet poles. Let $D(z_1, \lambda) = D(z_2, \lambda) = 0$ where $z_1 \neq z_2$ and $z_1 \neq z_2^*$. By eqn (4.36),

$$\frac{1}{z_i^2} = -\int\limits_{p_1}^{p_2} \frac{\sigma(z')}{z'+z_i}\,dz' \quad (i = 1, 2), \tag{4.45}$$

where
$$\sigma(z') = \frac{\lambda}{4\pi} z'\bar{\rho}(z')D(z', \lambda).$$

Clearly $\sigma(z) \geqslant 0$ on $p_1 \leqslant z \leqslant p_2$. Taking the difference of the two eqns (4.45) gives

$$\frac{1}{z_1 z_2}\left(\frac{1}{z_1}+\frac{1}{z_2}\right) = -\int_{p_1}^{p_2} \frac{\sigma(z')}{(z'+z_1)(z'+z_2)}\,dz'.$$

Using eqn (4.45) for the left-hand side gives

$$\int_{p_1}^{p_2} dz'\sigma(z')\left\{\frac{1}{(z'+z_1)(z'+z_2)}-\frac{1}{z_2(z'+z_1)}-\frac{1}{z_1(z'+z_2)}\right\} = 0,$$

i.e.
$$\int_{p_1}^{p_2} dz'\,\frac{z'^2\sigma(z')}{(z'+z_1)(z'+z_2)} = \int_{p_1}^{p_2} dz'\sigma(z'). \tag{4.46}$$

Equation (4.46) will also hold if z_2 is replaced by z_2^*, since $D(z_2^*, \lambda) = 0$. Therefore

$$(\operatorname{im} z_2)\int_{p_1}^{p_2} dz'\,\frac{z'^2\sigma(z')}{(z'+z_1)|z'+z_2|^2} = 0.$$

This is impossible if $\operatorname{im} z_1 \neq 0$, $\operatorname{im} z_2 \neq 0$, since

$$\operatorname{im}\int_{p_1}^{p_2} dz'\,\frac{z'^2\sigma(z')}{(z'+z_1)|z'+z_2|^2} = -(\operatorname{im} z_1)\int_{p_1}^{p_2} dz'\,\frac{z'^2\sigma(z')}{|z'+z_1|^2|z'+z_2|^2} \neq 0.$$

So $D(z, \lambda)$ can have at most one complex conjugate pair of zeros off the real axis.

4.7. Properties of the second sheet complex poles

Again with the same form for $\rho(s)$ we introduce the positive numbers b_i:

$$b_i = \frac{\lambda}{4\pi}\int_{p_1}^{p_2} (z')^i\bar{\rho}(z')D(z', \lambda)\,dz' \quad (i = 1, 2, \ldots), \tag{4.47}$$

where $0 < \lambda < \lambda_1$. The identity

$$\frac{1}{|z'+z|^2} = \frac{1}{|z|^2}-\frac{z'^2+2z'\operatorname{re} z}{|z|^2|z'+z|^2}$$

relates the numbers b_i to the positive functions $a_i(z)$ of eqn (4.37),

$$|z|^2a_i(z)+2\operatorname{re} z\,.\,a_{i+1}(z)+a_{i+2}(z) = b_i. \tag{4.48}$$

Therefore the branch of $\operatorname{im} D(z, \lambda) = 0$ which is given by eqn (4.41) can be written

$$a_3(z) = b_1. \tag{4.49}$$

Let $D(\tilde{z}, \lambda) = 0$ where $\operatorname{im}\tilde{z} \neq 0$. Then $\tilde{z}$ obeys eqn (4.49). By eqn (4.39)

$$\begin{aligned} 0 &= \operatorname{re} D(\tilde{z}, \lambda) \\ &= 1+a_1(\tilde{z})|\tilde{z}|^3\cos\tilde{\psi}+a_2(\tilde{z})|\tilde{z}|^2\cos(2\tilde{\psi}), \end{aligned} \tag{4.50}$$

where $\tilde{z} = |\tilde{z}|\exp(i\tilde{\psi})$. Also $\tilde{z}$ obeys eqn (4.41), so eqn (4.50) gives

$$|\tilde{z}|^2 a_2(\tilde{z}) = 1. \tag{4.51}$$

Combining this with eqn (4.49) we can write

$$|\tilde{z}|^2 = \frac{1}{b_1}\frac{a_3(\tilde{z})}{a_2(\tilde{z})}, \tag{4.52}$$

and now eqn (4.38) gives

$$\frac{p_1}{b_1} \leqslant |\tilde{z}|^2 \leqslant \frac{p_2}{b_1}. \tag{4.53}$$

By Appendix III and eqn (4.47), b_1 increases monotonically with λ in $0 < \lambda < \lambda_1$ and $b_1 \to +\infty$ as $\lambda \to \lambda_1$. Therefore $|\tilde{z}| \to 0$ as $\lambda \to \lambda_1$. Equation (4.53) shows that for a complex pole, $|\tilde{z}|$ lies between two monotonically decreasing bounds as λ increases, but it does not prove that $|\tilde{z}|$ decreases monotonically (in the single pole case eqn (4.19) showed that $|\tilde{z}|$ decreases monotonically as γ increases from $\frac{2}{9}$ to 2).

4.8. Form of the curve im $D(z,\lambda) = 0$

Since $D(z^*) = D^*(z)$, it is only necessary to study the region $\operatorname{im} z \geqslant 0$. We saw in § 4.5 and § 4.7 (especially eqn (4.49)) that the branches of the curve $\operatorname{im} D(z,\lambda) = 0$ (for fixed λ in $0 < \lambda < \lambda_1$) are the curve $a_3(z) = b_1$, and the real axis except for the set of points E on $-p_2 \leqslant z \leqslant -p_1$.

Equation (4.37) shows that $a_3(z)$ decreases monotonically as z moves away from the real axis along any line $\operatorname{re} z = \text{constant}$. Such a line can cross $a_3(z) = b_1$ at most once in $\operatorname{im} z > 0$.

Let $-\infty < x < 0$ and suppose x is in the set E. Then

$$a_3(x+iy) \to +\infty, \quad \text{as } y \to 0;$$
$$a_3(x+iy) \to 0, \quad \text{as } y \to +\infty.$$

Thus the line $\operatorname{re} z = x$ must intersect the curve $a_3(z) = b_1$ once in $\operatorname{im} z > 0$.

Let $-\infty < x < 0$ and suppose x is not in the set E. The line $\operatorname{re} z = x$ will intersect $a_3(z) = b_1$ in $\operatorname{im} z > 0$ if, and only if, $a_3(x) > b_1$. If $a_3(x) = b_1$ and $D(z,\lambda)$ is regular at $z = x$, by eqn (4.37)

$$a_3(x+iy) = b_1 + O(y^2),$$

and the curve $a_3(z) = b_1$ must meet the real axis at x at right angles (compare with section 3.3).

On moving along the real axis $-\infty \leqslant x \leqslant -p_2$ from $-\infty$ to $-p_2$, $a_3(x)$ increases monotonically from zero, and either $a_3(x) \to +\infty$ or $a_3(x) \to a_3(-p_2)$ as $x \to -p_2$. The finite value can only occur if $\bar{\rho}(p_2) = 0$ and $\bar{\rho}(p_2 - \epsilon)$ goes to zero sufficiently fast as $\epsilon \to 0$ (see eqn (4.37)). We shall now write $a_3(-p_2)$ for $+\infty$ or the finite value.

If $a_3(-p_2) > b_1$, there is a unique point z_c $(-\infty < z_c < -p_2)$ for which $a_3(z_c) = b_1$. For $-\infty < x < z_c$ the line $\mathrm{re}\, z = x$ cannot cross $a_3(z) = b_1$, and for $z_c < x < -p_2$ the line $\mathrm{re}\, z = x$ will cross $a_3(z) = b_1$ in $\mathrm{im}\, z > 0$.

If $a_3(-p_2) \leqslant b_1$, then $a_3(x) < b_1$ for $-\infty \leqslant x < -p_2$, so $\mathrm{re}\, z = x$ will not cross $a_3(z) = b_1$, for $-\infty \leqslant x < -p_2$. To see what happens at $-p_2$, we can choose a sequence of points $\{x_n\}$ in the set E such that $x_n \to -p_2$ as $n \to \infty$. So $\mathrm{re}\, z = x_n$ crosses $a_3(z) = b_1$, for all x_n. Therefore a branch of $\mathrm{im}\, D(z, \lambda) = 0$ must leave the real axis at $z = -p_2$ and must bend off to the right of the line $\mathrm{re}\, z = -p_2$ in $\mathrm{im}\, z \neq 0$. In this case we write $z_c = -p_2$.

By eqns (4.42) a branch of $\mathrm{im}\, D(z, \lambda) = 0$ can only leave the real axis at $z = -p_2$ if $p_2 \geqslant 2p_1$. Thus $a_3(-p_2) \leqslant b_1$ only if $p_2 \geqslant 2p_1$. It can be shown that if $p_2 > 2p_1$ and if $a_3(-p_2) > b_1$ for small λ then $a_3(-p_2) > b_1$ for all λ $(0 < \lambda < \lambda_1)$.

Typical cases of the behaviour of the curve $\mathrm{im}\, D(z, \lambda) = 0$ are shown in Figs. 4.6 (a) and 4.6 (b).

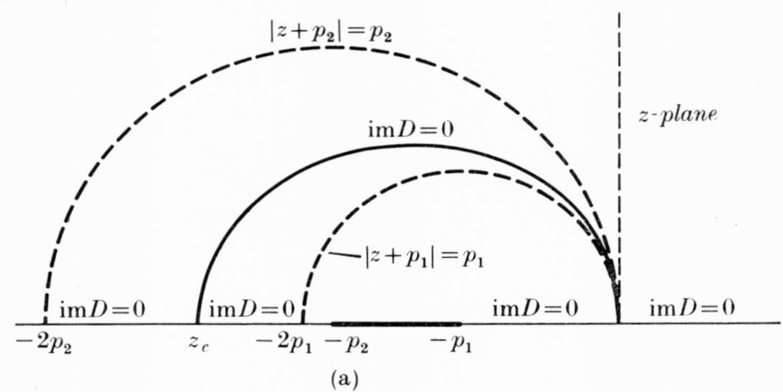

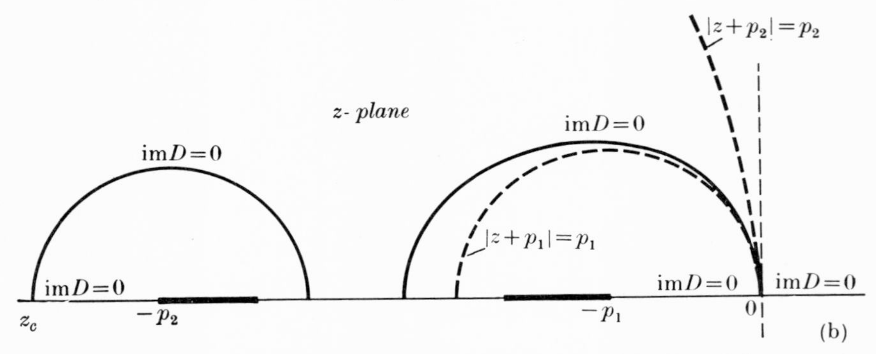

FIG. 4.6. (a) The solid curve shows a possible behaviour of $\mathrm{im}\, D = 0$ for $p_2 < 2p_1$. (b) The solid curve shows a possible behaviour of $\mathrm{im}\, D = 0$ for $p_2 > 2p_1$.

4.9. Behaviour of re $D(z, \lambda)$ and the motion of the poles

It is clear from eqn (4.40) that for large $|z|$,

$$\operatorname{im} D(z, \lambda) \simeq b_1 |z| \sin \psi.$$

Let C^+ denote the branch of $\operatorname{im} D = 0$ which (for fixed λ) is the boundary of the region in $\operatorname{im} z \geqslant 0$ where $\operatorname{im} D > 0$. An example is shown in Fig. 4.7. Moving from $-\infty$ to $+\infty$ along C^+, the region $\operatorname{im} D > 0$ is always on the left. From the general discussion in Chapter 3 this shows that $\operatorname{re} D$ increases monotonically on C^+. Therefore there can be at most one zero of $D(z, \lambda)$ on C^+ (this contains the result of § 4.6).

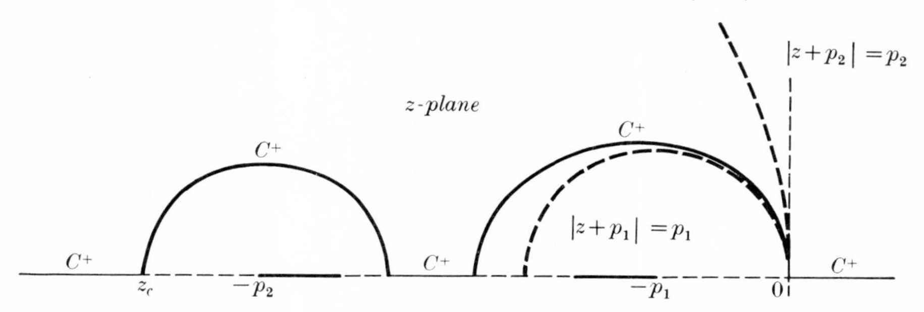

Fig. 4.7. The solid line is the branch C^+ of $\operatorname{im} D(z, \lambda) = 0$.

Since $D(0, \lambda) = 1$, and by eqn (4.36) $\operatorname{re} D(x, \lambda) \to -\infty$ as $x \to -\infty$, there is one zero of $D(z, \lambda)$ on C^+ in $\operatorname{re} z < 0$. Let this zero be $\tilde{z}(\lambda)$. We now examine how $\tilde{z}(\lambda)$ moves as λ increases from 0 to λ_1.

Poles on the real axis

For real x, by eqn (4.36), $\operatorname{re} D(x, \lambda) = 0$ can be written

$$R(x, \lambda) = -1, \tag{4.54}$$

where

$$R(x, \lambda) = \frac{\lambda}{4\pi} x^2 \int\limits_{p_1}^{p_2} \frac{z'\bar{\rho}(z')D(z', \lambda)}{z'+x}\, dz'. \tag{4.55}$$

For fixed λ, $\operatorname{re} D(x, \lambda) = 1 + R(x, \lambda)$ increases monotonically on each segment of C^+ which lies on the real axis. The same holds for $R(x, \lambda)$. Let λ', λ'' be two values of λ in $0 < \lambda' < \lambda'' < \lambda_1$ for which $\tilde{z}(\lambda')$ and $\tilde{z}(\lambda'')$ lie on the real axis. By eqn (4.55) and the results of Appendix III,

$$R(\tilde{z}(\lambda'), \lambda'') < R(\tilde{z}(\lambda'), \lambda') = -1.$$

But $R(\tilde{z}(\lambda''), \lambda'') = -1$, therefore

$$\tilde{z}(\lambda') < \tilde{z}(\lambda''). \tag{4.56}$$

Thus when $\tilde{z}(\lambda)$ is real it increases monotonically with λ.

For $-\infty < x < -p_2$, eqn (4.55) gives

$$\frac{x^2 b_1}{x+p_2} \leqslant R(x,\lambda) \leqslant \frac{x^2 b_1}{x+p_1}. \tag{4.57}$$

Now

$$\frac{x^2 b_1}{x+p_i} = -1$$

for

$$x = -\frac{1}{2b_1}\{1+\sqrt{(1-4b_1 p_i)}\} \quad (i = 1, 2).$$

Taking the other sign of the square roots gives a segment in which eqn (4.57) cannot be satisfied, but with the sign given here, and by eqns (4.54) and (4.57),

$$-\frac{1}{2b_1}\{1+\sqrt{(1-4b_1 p_1)}\} \leqslant \tilde{z}(\lambda) \leqslant -\frac{1}{2b_1}\{1+\sqrt{(1-4b_1 p_2)}\}, \tag{4.58}$$

provided that $-\{1+\sqrt{(1-4b_1 p_i)}\}/2b_1$ $(i = 1, 2)$ are real and less than $-p_2$. This is the case for

$$b_1 < 1/4p_2. \tag{4.59}$$

By eqn (4.47) $b_1 \to 0$ as $\lambda \to 0$. Therefore by eqns (4.58) and (4.59) $\tilde{z}(\lambda) \to -\infty$ as $\lambda \to 0$. Hence $\tilde{z}(\lambda)$ increases monotonically from $-\infty$ as λ increases from zero. We saw in § 4.8 that $z_c(\lambda) \leqslant -p_2$. Therefore for some value λ_c $(0 < \lambda_c < \lambda_1)$, $\tilde{z}(\lambda_c)$ coincides with $z_c(\lambda_c)$. As λ increases above λ_c, $\tilde{z}(\lambda)$ must leave the real axis.

For $z_c < -p_2$ we have $\text{im}\, D(z,\lambda) < 0$ above the interval $z_c < x < -p_2$, so $\text{re}\, D(x,\lambda)$ decreases monotonically as x moves from z_c to $-p_2$. There can be at most one zero of $\text{re}\, D(x,\lambda)$ on this interval.

By eqn (4.55), $R(x,\lambda) < 0$ for $x \leqslant -p_2$ and $0 < \lambda < \lambda_1$, so for fixed $x \leqslant -p_2$, $R(x,\lambda)$ will decrease as λ increases. It may happen that $R(x,\lambda) \to -\infty$ as $x \to -p_2$ (in particular this is the case if $\bar{\rho}(z) \neq 0$ at, and near, p_2). Always

$$\lim_{x \to -p_2} R(x, 0) = 0.$$

Let λ_c' be the largest value of λ for which

$$\lim_{x \to -p_2} R(x,\lambda) \geqslant -1.$$

Then $0 \leqslant \lambda_c' \leqslant \lambda_c$, since

$$\lim_{x \to -p_2} R(x,\lambda_c) \leqslant R(z_c(\lambda_c),\lambda_c) = -1.$$

As λ increases from λ_c' to λ_c a zero of $D(z,\lambda)$ moves monotonically from $-p_2$ to $z_c(\lambda_c)$.

For $\lambda = \lambda_c$ two zeros of $D(z,\lambda)$ coincide at $z_c(\lambda_c)$. The positions of the zeros $\tilde{z}(\lambda)$ are analytic functions of λ, so for λ near λ_c we must have

$$\tilde{z}(\lambda) = z_c(\lambda_c) + k(\lambda_c - \lambda)^{\frac{1}{2}},$$

where k is a constant. Hence if λ moves above λ_c the positions $\tilde{z}(\lambda)$ move off the real axis and their paths are initially tangential to the line $\operatorname{re} z = z_c(\lambda_c)$ (this argument may not hold if $z_c(\lambda_c) = -p_2$). The two positions are then complex conjugates.

Also, for $\lambda > \lambda_c$, $x \leqslant -p_2$,

$$R(x, \lambda) < R(x, \lambda_c) \leqslant -1,$$

so there cannot be any zeros of $D(z, \lambda)$ on $-\infty < x < -p_2$ for $\lambda > \lambda_c$.

Complex poles

For $\lambda > \lambda_c$, if $\tilde{z}(\lambda)$ is complex we have eqn (4.53):

$$p_1/b_1 \leqslant |\tilde{z}(\lambda)|^2 \leqslant p_2/b_1.$$

The upper and the lower bounds in this equation decrease monotonically as λ increases. However, this does not exclude the possibility that $\tilde{z}(\lambda)$ returns to the real axis for some value $\lambda > \lambda_c$. By the immediately preceding argument $\tilde{z}(\lambda)$ cannot return to the segment $z_c(\lambda_c) \leqslant x \leqslant -p_2$. By eqn (4.51) and the argument in § 4.8, $\tilde{z}(\lambda)$ cannot return to any segment of the real axis where $\bar{\rho}(x) \neq 0$ (or more precisely, to any point of the set E on $-p_2 \leqslant z \leqslant -p_1$).

However, if $p_2 > 2p_1$, and if there is a segment of (p_1, p_2) where $\bar{\rho}(x) = 0$, then it may happen that as λ increases ($\lambda > \lambda_c$), $\tilde{z}(\lambda)$ returns to the real axis. That is, it may be that there is one (or perhaps several) interval(s) (x', x'') such that $-p_2 < x' < x'' < -2p_1$ and $\bar{\rho}(x) = 0$ on $x' \leqslant x \leqslant x''$, where $\tilde{z}(\lambda)$ reaches the real axis at x' and leaves it at x''. Actually $\tilde{z}(\lambda)$ and $\tilde{z}(\lambda)^*$ will reach the real axis at x' and leave it at x''.

On $x' \leqslant x \leqslant x''$, one pole position, $\tilde{z}(\lambda)$, will increase monotonically from x' to x'' as λ increases. The other pole will start from x' in the negative direction and reach x'' moving in the negative direction. These two pole positions can only meet at x' and x''.

By eqn (4.53) x'' must obey

$$p_1/b_1 \leqslant x''^2 \leqslant p_2/b_1$$

when $\tilde{z}(\lambda) = x''$. Therefore when λ has increased to a value such that

$$b_1 > p_2/x''^2,$$

$\tilde{z}(\lambda)$ must have left the interval $x' \leqslant x \leqslant x''$, and it cannot return to that interval (in $\lambda_c < \lambda < \lambda_1$). In particular for $b_1 > p_2/4p_1^2$, $\tilde{z}(\lambda)$ must be complex; it moves on a path similar to that in the single pole case (§ 4.2), and $|\tilde{z}(\lambda)| \to 0$ as $\lambda \to \lambda_1$.

Finally we notice that $D(z, \lambda)$ cannot vanish on $(-p_1, 0)$ since by eqn (4.55) $R(x, \lambda) > 0$ there provided $0 < \lambda < \lambda_1$.

4.10. The resonant energy

We *define* the resonant energy E_R by the condition

$$\alpha(s_R+) = \tfrac{1}{2}\pi,$$

where $s_R = E_R^2$ and $s_0 \leqslant s_R \leqslant \infty$. The effective range identity

$$q^3 \cot \alpha(s) = \operatorname{re} D(s)/N(s), \quad \text{for } s_0 \leqslant s \leqslant \infty,$$

shows that s_R is defined by $\operatorname{re} D(s_R) = 0$.

In the z-plane the point s_R+ becomes $-iz_R$ where $0 \leqslant z_R \leqslant \infty$. Thus the resonant energy is given by

$$\operatorname{re} D(-iz_R, \lambda) = 0, \quad \text{for } 0 \leqslant z_R \leqslant \infty. \tag{4.60}$$

By eqn (4.36) we can write eqn (4.60) as

$$\begin{aligned} z_R^2 a_2(-iz_R) &\equiv z_R^2 \frac{\lambda}{4\pi} \int\limits_{p_1}^{p_2} \frac{z'^2 \bar{\rho}(z') D(z', \lambda)}{z'^2 + z_R^2}\, dz' \\ &= 1. \end{aligned} \tag{4.61}$$

Using eqn (4.48) we can alternatively write

$$\begin{aligned} a_4(-iz_R) &\equiv \frac{\lambda}{4\pi} \int\limits_{p_1}^{p_2} \frac{z'^4 \bar{\rho}(z') D(z', \lambda)}{z'^2 + z_R^2}\, dz' \\ &= b_2 - 1. \end{aligned} \tag{4.62}$$

For real positive y, $y^2 a_2(-iy)$ increases monotonically with y for fixed λ. Thus eqn (4.61) has at most one solution. Thus if $\alpha(s+)$ does increase through $\frac{1}{2}\pi$ as s increases ($s_0 \leqslant s \leqslant \infty$), it cannot again pass through $\frac{1}{2}\pi$. Also for fixed y, $y^2 a_2(-iy)$ increases monotonically with λ. Therefore a solution z_R of eqn (4.61) must decrease monotonically to zero as λ increases to λ_1. By eqn (4.47) $b_2(\lambda)$ increases monotonically from 0 to ∞ as λ increases from 0 to λ_1. Let $b_2(\lambda_R) = 1$. For $\lambda < \lambda_R$, $b_2(\lambda) < 1$, and eqn (4.62) has no solution for real z_R. For $\lambda = \lambda_R$ eqn (4.62) has the solution $z_R \to \infty$. As λ increases above λ_R, the solution z_R decreases monotonically from $+\infty$ to zero. Thus for $\lambda < \lambda_R$ there is no resonant energy, and for $\lambda_R < \lambda < \lambda_1$ there is a resonant energy s_R which decreases monotonically from $+\infty$ to s_0 as λ increases in this range.

Further restriction on the resonance poles

By eqns (4.48) and (4.51) the pole position $\tilde{z}$ obeys

$$2a_3(\tilde{z}) . \operatorname{re} \tilde{z} = b_2 - 1 - a_4(\tilde{z}). \tag{4.63}$$

By eqn (4.38)

$$p_1 \leqslant \frac{a_4(\tilde{z})}{a_3(\tilde{z})} \leqslant p_2.$$

Therefore

$$\operatorname{re} \tilde{z} \leqslant -\tfrac{1}{2}p_1, \quad \text{for } \lambda \leqslant \lambda_R; \tag{4.64}$$

$$\operatorname{re} \tilde{z} \geqslant -\tfrac{1}{2}p_2, \quad \text{for } \lambda \geqslant \lambda_R. \tag{4.65}$$

It follows from eqn (4.65) that $\lambda_R > \lambda_c$ since $z_c(\lambda_c) < -p_2$. Equations (4.64) and (4.65) give us information on the location of the second sheet poles corresponding to resonant amplitudes (Fig. 4.8). In particular such poles cannot lie on the real axis ($-\infty < s < s_0$) if $p_2 < 4p_1$.

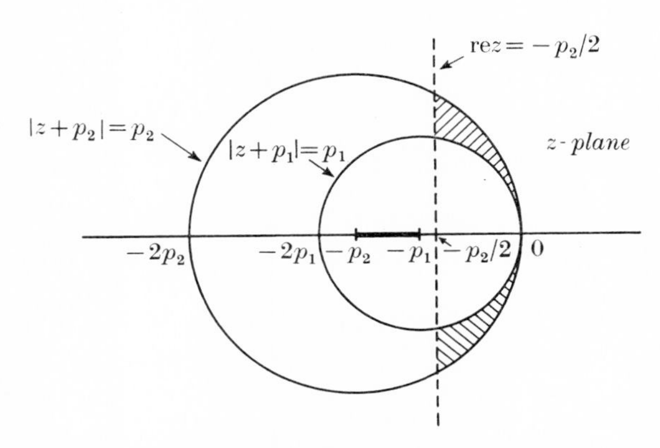

FIG. 4.8. The shaded region is where the resonance poles can lie.

Since these resonance poles cannot lie in the region $\operatorname{re} s > s_0 + \tfrac{1}{2}p_2^2$ there is a range of values of λ ($\lambda > \lambda_R$) for which the distance between s_R and the resonance poles is large. In these cases the resonance will be very broad and will have an abnornal shape. For $\lambda \to \lambda_1$ the distance between the resonance poles and s_R goes to zero. Also if $s_R - s_0 = O(p_1^2)$ then s_R will not be far from the poles.

4.11. *P*-wave with weak inelasticity

In the real world the p.w.a. will not be elastic and there can be numerous unphysical sheets opening at the various thresholds. The complete description of the many channel amplitudes will involve studying the poles on the several unphysical sheets.

It may, however, be useful, whenever the inelasticity is not strong, to examine the second sheet poles of the equivalent elastic amplitude which was defined in § 2.3. The inelasticity parameter $\eta(s)$ is specified, and it is assumed to be independent of the parameter λ which we use to give the strength of the interaction, although it may not be a good physical assumption to take $\eta(s)$ independent of λ.

We shall examine briefly some of the changes required in the discussion from § 4.4 onwards. The equivalent elastic reduced p.w.a. $\tilde{F}(s)$ is studied. It is defined by

$$2i(q(s))^3\tilde{F}(s)+1 = L(s)\{2i(q(s))^3F(s)+1\}, \tag{4.66}$$

where $F(s)$ is the reduced P-wave amplitude and $L(s)$ is given by eqn (2.20). The equivalent elastic phase is

$$\tilde{\alpha}(s) = \alpha(s)+\tfrac{1}{2}\phi(s), \tag{4.67}$$

where $\alpha(s)$ is the real phase and $\phi(s)$ is defined in eqn (2.22).

We write
$$L(s) = \exp\{i(s-s_0)^{\frac{1}{2}}\chi(s)\}, \tag{4.68}$$

where
$$\chi(s) = \frac{1}{\pi}\int\limits_{s_0}^{\infty} ds' \frac{\ln \eta(s')}{(s'-s_0)^{\frac{1}{2}}(s'-s)}. \tag{4.69}$$

On the physical cut near s_0, $\eta(s) = 1$, so $\chi(s)$ is regular near s_0. Equation (4.66) gives

$$\tilde{F}(s) = L(s)F(s)+\frac{1}{2i(q(s))^3}(L(s)-1). \tag{4.70}$$

Since $\chi(s_0) < 0$ if there is any inelasticity, it follows from eqn (4.70) that $\tilde{F}(s)$ has a pole at s_0. By eqn (4.1) this pole is

$$\frac{4\chi(s_0)}{s-s_0}. \tag{4.71}$$

This pole arises because $\phi(s) = (s-s_0)^{\frac{1}{2}}\chi(s_0)$ near s_0, so by eqn (4.67) $\tilde{\alpha}(s)$ behaves like $(s-s_0)^{\frac{1}{2}}$, rather than $(s-s_0)^{\frac{3}{2}}$, near s_0.

The difficulty is surmounted by modifying the N/D equations. We write $\tilde{F} = N/D$, and, as usual, we require $D(s_0) = 1$. We write

$$N(s) = \frac{4\chi(s_0)}{s-s_0}+N'(s), \tag{4.72}$$

where $N'(s)$ is regular except for the left-hand cut $s_2 \leqslant s \leqslant s_1$, where

$$\operatorname{im} N'(s+) = \tilde{\rho}(s)D(s) \tag{4.73}$$

and
$$\tilde{\rho}(s) = L(s)\operatorname{im} F(s+) \quad (s_2 \leqslant s \leqslant s_1).$$

Also on $s_0 \leqslant s \leqslant \infty$,

$$\begin{aligned}\operatorname{im} D(s+) &= -q^3N(s)\\ &= -\tfrac{1}{2}(s-s_0)^{\frac{1}{2}}\chi(s_0)-q^3N'(s).\end{aligned} \tag{4.74}$$

Now eqns (4.72), (4.73), and (4.74) are used in § 4.4. The only differences from the results stated there are that eqn (4.33) is replaced by the relation for small $|z|$,

$$D(z) = 1+\tfrac{1}{2}\chi(s_0)z+c_2z^2+c_3z^3+\dots, \tag{4.75}$$

and we write the dispersion relation for the function

$$(D(z)-1-\tfrac{1}{2}\chi(s_0)z)/z^2,$$

giving, in place of eqn (4.35),

$$D(z) = 1+\tfrac{1}{2}\chi(s_0)z-\frac{z^2}{4\pi}\int_{p_1}^{p_2}\frac{z'\tilde{\rho}(s_0-z'^2)D(z')}{z'+z}\,dz'. \tag{4.76}$$

Locating the poles

We wish to find the zeros of $D(z)$. We write

$$\tilde{\rho}(s) = -\lambda\bar{\rho}(z)$$

and

$$c = \tfrac{1}{2}\chi(s_0). \tag{4.77}$$

Then eqn (4.76) becomes

$$D(z,\lambda) = 1+cz+\int_{p_1}^{p_2}\frac{z^2\sigma(z')}{z+z'}\,dz', \tag{4.78}$$

where

$$\sigma(z) = \frac{\lambda}{4\pi}z\bar{\rho}(z)D(z,\lambda). \tag{4.79}$$

For the attractive interaction, $\bar{\rho}(z) \geqslant 0$ $(p_1 \leqslant z \leqslant p_2)$. By eqns (4.69) and (4.77), $c < 0$ if there is any inelasticity. We shall now assume that

$$1+cp_2 > 0,$$

which limits the amount of inelasticity we can allow in this treatment.

It follows that $D(z,0) > 0$ on $p_1 \leqslant z \leqslant p_2$. By arguments similar to those in Appendix III it can be shown that for fixed z in $p_1 \leqslant z \leqslant p_2$, $D(z,\lambda)$ is a monotonic increasing function of λ for $0 < \lambda < \lambda_1$, and $D(z,\lambda) \to +\infty$ as $\lambda \to \lambda_1$. Hence $D(z,\lambda) > 0$ for $p_1 \leqslant z \leqslant p_2$, $0 \leqslant \lambda < \lambda_1$. It follows that $\sigma(z) \geqslant 0$ for $p_1 \leqslant z \leqslant p_2$.

Now, by eqn (4.78),

$$\operatorname{im}\{(z+p_2)z^{*2}D(z,\lambda)\}$$

$$= (\operatorname{im} z^*)\left\{(1+cp_2)|z|^2+2p_2\operatorname{re} z+|z|^4\int_{p_1}^{p_2}\frac{p_2-z'}{|z+z'|^2}\sigma(z')\,dz'\right\}. \tag{4.80}$$

The last factor on the right of eqn (4.80) is positive for

$$(1+cp_2)|z|^2+2p_2\operatorname{re} z > 0,$$

i.e. for

$$|z| > -2\frac{p_2}{1+cp_2}\cos\psi, \tag{4.81}$$

where $z = |z|\exp(i\psi)$. So $D(z,\lambda)$ cannot have a zero in the region where eqn (4.81) holds and $\operatorname{im} z \neq 0$ (and $0 < \lambda < \lambda_1$). Similarly, using

$$\operatorname{im}\{(z+p_1)z^{*2}D(z,\lambda)\},$$

it follows that $D(z, \lambda)$ cannot have a zero in the region where

$$|z| < -2\frac{p_1}{1+cp_1}\cos\psi \tag{4.82}$$

and $\operatorname{im} z \neq 0$.

Thus for $0 < \lambda < \lambda_1$, the zeros of $D(z, \lambda)$ lying off the real axis must lie in the region between the circles

$$|z| = -\frac{2p_2}{1+cp_2}\cos\psi \tag{4.83}$$

and

$$|z| = -\frac{2p_1}{1+cp_1}\cos\psi. \tag{4.84}$$

This region lies in $\operatorname{re} z < 0$ (except for $z = 0$), therefore such zeros of $D(z, \lambda)$ give the complex second sheet poles of $\tilde{F}(s)$.

When there is a single driving pole, $p_1 = p_2$ and the two circles in eqns (4.83) and (4.84) coincide. Then they give the pole curve. It follows that eqns (4.83) and (4.84) are the optimum which can be obtained unless the discontinuity $\rho(s)$ is specified further.

Motion of the complex poles

We use the positive real functions $a_i(z)$ defined in eqn (4.37). Then eqn (4.78) can be written

$$D(z, \lambda) = 1+cz+z^2a_2(z)+z|z|^2a_1(z). \tag{4.85}$$

The branches of $\operatorname{im} D(z, \lambda) = 0$ are thus (a) the real axis, $\sin\psi = 0$, and (b) the curve

$$c+2a_2(z)|z|\cos\psi+a_1(z)|z|^2 = 0. \tag{4.86}$$

Using eqn (4.48), we can write eqn (4.86) as

$$a_3(z) = c+b_1. \tag{4.87}$$

Now b_1 increases monotonically from 0 to $+\infty$ as λ goes from 0 to λ_1. Let λ_b be the value which satisfies

$$b_1(\lambda_b)+c = 0. \tag{4.88}$$

It follows that a zero of $D(z, \lambda)$ cannot lie off the real axis unless $\lambda_b < \lambda < \lambda_1$.

By eqn (4.85),

$$\operatorname{re} D(z, \lambda) = 1+c|z|\cos\psi+|z|^2a_2(z)\cos(2\psi)+|z|^3a_1(z)\cos\psi. \tag{4.89}$$

Let $\tilde{z}$ be a zero of $D(z, \lambda)$ which does not lie on the real axis. Then, by eqns (4.86) and (4.89),

$$|\tilde{z}|^2a_2(\tilde{z}) = 1. \tag{4.90}$$

Now using eqns (4.38) and (4.87) we see that a complex zero of $D(z, \lambda)$ must obey

$$\frac{p_1}{c+b_1} \leqslant |\tilde{z}|^2 \leqslant \frac{p_2}{c+b_1}, \tag{4.91}$$

and as $\lambda \to \lambda_1$, $|\tilde{z}|$ must go to zero.

Using these equations and arguments similar to those in the elastic case it is easy to get the following results:

(1) For $0 \leqslant \lambda \leqslant \lambda_b$, there are no complex zeros of $D(z, \lambda)$. There is one and only one zero on $-p_1 \leqslant z \leqslant \infty$. As λ increases from 0 to λ_b this zero moves monotonically from $-c^{-1}$ to $+\infty$. Since $-c^{-1} > p_2$ this gives a physical sheet pole on $-\infty \leqslant s \leqslant s_2$. On $-\infty \leqslant z \leqslant -p_2$ there is at most one zero of $D(z, \lambda)$.

(2) For $\lambda_b \leqslant \lambda \leqslant \lambda_1$, the motion of the poles is very similar to the elastic case. Arguments like those in §§ 4.8 and 4.9 show that there can be at most one pair of complex conjugate zeros of $D(z, \lambda)$. As λ increases from λ_b a zero moves monotonically along the negative real axis from $-\infty$ to a point z_c where $z_c \leqslant -p_2$; it leaves the real axis at z_c. If $p_2 > 2p_1/(1+cp_1)$ the zero may return to the real axis in the interval $(-p_2, -2p_1/(1+cp_1))$, but as λ increases it must leave this interval again before

$$\frac{p_2}{c+b_1} < \left(\frac{2p_1}{1+cp_1}\right)^2.$$

The pair of complex zeros must lie between the circles in eqns (4.83) and (4.84), and as $\lambda \to \lambda_1$, they will meet at $z = 0$.

5

POLE REGIONS FOR ATTRACTIVE *P*-WAVES: PION–PION SCATTERING

5.1. Introduction

WE now examine the properties of the isolated elastic P-wave solutions for π–π scattering. As was explained in Chapter 1 it is appropriate in this case to specify $\operatorname{im} A(s+)$ on the left-hand cut. Because of the extra kinematic factor in $A(s)$ the method and the results are somewhat different from those for the p.w.a. $f(s)$ which we studied in Chapter 4. In place of the rational functions appearing in Chapter 4 we have here circular and logarithmic functions, so the manipulations are a little more complicated.

The general form of the results on the location of the second sheet poles arising from a truncated attractive left-hand cut are the same as those in Chapter 4, but there are important differences in detail. For example, there can be more than one pair of poles off the real axis; another difference is that the resonance energy s_R cannot be arbitrarily large if the left-hand cut is of fixed length and position.

We should also warn the reader that it has been found convenient to make a few changes in notation compared with Chapter 4, for example the definition of z (eqn (5.7)).

5.2. The integral equation for $D(z)$

Notation

We have
$$s = 4(q^2+\mu^2),$$
where μ is the pion's mass, and q is the momentum in the c.m.s. It is often convenient to use the variable
$$\nu = q^2$$
in place of s. The physical threshold is $s_0 = 4\mu^2$ (i.e. $\nu = 0$). The scattering amplitude is
$$A(s) = \frac{(q^2+\mu^2)^{\frac{1}{2}}}{\mu}\frac{f_1(s)}{q^2}, \tag{5.1}$$
where $f_1(s)$ is the ordinary P-wave p.w.a. We only deal with elastic scattering. In general the left-hand cut of $A(s)$ is $-\infty \leqslant s \leqslant 0$, and we

shall assume that we know

$$\rho(s) = \operatorname{im} A(s+) \quad (-\infty \leqslant s \leqslant 0),$$

on the left-hand cut.

Since we are dealing with the reduced amplitude, by § 2.5 we can write the dispersion relation,

$$A(s) = \frac{1}{\pi}\int\limits_{-\infty}^{0} \frac{\rho(s')\,\mathrm{d}s'}{s'-s} + \frac{1}{\pi}\int\limits_{s_0}^{\infty} \frac{\operatorname{im} A(s'+)}{s'-s}\,\mathrm{d}s', \tag{5.2}$$

and its isolated solution is given by the solution of the N/D equations

$$A(s) = N(\nu)/D(\nu), \tag{5.3}$$

$$N(\nu) = \frac{1}{\pi}\int\limits_{-\infty}^{-1} \mathrm{d}\nu' \, \frac{\rho(\nu')D(\nu')}{\nu'-\nu}, \tag{5.4}$$

$$D(\nu) = 1 - \frac{\nu}{\pi}\int\limits_{0}^{\infty} \mathrm{d}\nu' \left(\frac{\nu'^3}{\nu'+1}\right)^{\frac{1}{2}} \frac{N(\nu')}{\nu'(\nu'-\nu)}. \tag{5.5}$$

In eqns (5.3), (5.4), and (5.5) we have for convenience put $\mu = 1$, and in place of $\rho(s)$ we have written $\rho(\nu)$. In eqn (5.5) the unitarity relation

$$\begin{aligned} \operatorname{im} D(\nu+) &= N(\nu)\operatorname{im}(A(s+))^{-1} \\ &= -\left(\frac{\nu^3}{\nu+1}\right)^{\frac{1}{2}} N(\nu), \quad \text{for } 0 \leqslant \nu \leqslant \infty, \end{aligned} \tag{5.6}$$

has been used. This comes from eqn (5.1) and the assumption of elasticity. In § 5.7 we show that the solution of eqns (5.3), (5.4), and (5.5) is actually the isolated solution.

In this chapter we define

$$z = -iq. \tag{5.7}$$

This differs by a factor 2 from the definition used in § 4.2 (cf. eqns (4.5). and (4.6) and Fig. 4.1) but it is otherwise the same. The physical s-sheet is $\operatorname{re} z \geqslant 0$, and the second s-sheet is $\operatorname{re} z \leqslant 0$.

We also require the function

$$\omega(z) = (1-z^2)^{\frac{1}{2}}. \tag{5.8}$$

It is defined in the z-plane cut along $-\infty \leqslant z \leqslant -1$ and $1 \leqslant z \leqslant \infty$, so that on the cuts it takes the values shown in Fig. 5.1. In this cut plane we have

$$\omega(-z) = \omega(z),$$

$$\omega(z^*) = \omega^*(z).$$

Also

$$\omega(0) = 1.$$

When we express N and D as functions of z we shall write them as $N(z)$ and $D(z)$. Using eqns (5.7) and (3.3) we have $z(s^*) = z^*(s)$, so the reality conditions are written

$$N^*(z) = N(z^*),$$
$$D^*(z) = D(z^*).$$

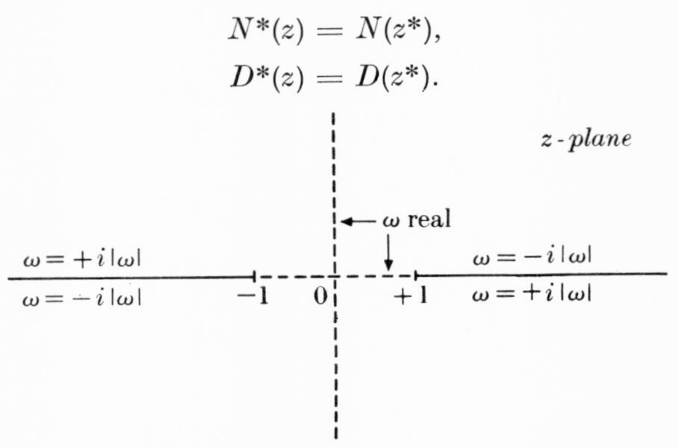

FIG. 5.1. The definition of $\omega(z)$ in the z-plane cut along $-\infty \leqslant z \leqslant -1$ and $1 \leqslant z \leqslant \infty$. On re $z = 0$, $\omega \geqslant 1$; and on $-1 \leqslant z \leqslant 1$, $0 \leqslant \omega \leqslant 1$.

Analytic continuation

Equations (5.4) and (5.5) define $N(z)$, $D(z)$ in $\operatorname{re} z \geqslant 0$, and we wish to continue these functions to $\operatorname{re} z < 0$. Equation (5.4) shows that $N(z)$ is a regular function of z^2 near $z = 0$; therefore

$$N(-z) = N(z) \tag{5.9}$$

in the z-plane cut along $-\infty \leqslant z \leqslant -1$, $1 \leqslant z \leqslant \infty$.

Let $$H(-z) = D(z)+2(z^3/\omega)N(z), \quad \text{for } \operatorname{re} z \geqslant 0. \tag{5.10}$$

The reality condition and eqn (5.6) show that on the physical axis† $\operatorname{re} z = 0$,

$$\operatorname{im} D(z) = \operatorname{sg}(z/i)\frac{|z|^3}{|1-z^2|^{\frac{1}{2}}}N(z),$$

where $\operatorname{sg}(x) = x/|x|$ for real x. Thus on $\operatorname{re} z = 0$,

$$\operatorname{im} H(-z) = -\operatorname{im} D(z),$$
$$\operatorname{re} H(-z) = \operatorname{re} D(z),$$

so $$H(-z) = D^*(z) = D(-z).$$

It follows, as in § 4.4, that eqn (5.10) gives

$$D(-z) = D(z)+2(z^3/\omega)N(z) \tag{5.11}$$

for any z in the plane cut along $-\infty \leqslant z \leqslant -1$, $1 \leqslant z \leqslant \infty$. Equation (5.5) shows that $D(z)$ is regular for $\operatorname{re} z > 0$, so $D(z)$ is regular in the z-plane cut along $-\infty \leqslant z \leqslant -1$.

† When we write $| \ldots |^{\frac{1}{2}}$ we always mean the positive root.

On $\operatorname{re} z = 0$, the function $S(z)$, defined by eqn (3.9), obeys

$$\begin{aligned} S(z) &= 1+2iq^3(z)(1+q^2(z))^{-\frac{1}{2}}A(z) \\ &= \{D(z)+2(z^3/\omega)N(z)\}/D(z) \\ &= D(-z)/D(z). \end{aligned} \tag{5.12}$$

Clearly eqn (5.12) is valid in the plane cut along $-\infty \leqslant z \leqslant -1$, $1 \leqslant z \leqslant \infty$.

Since $N(z)$ and $D(z)$ are regular in $|z| < 1$, the unitarity relation (eqn (5.6)) shows that the Taylor series for $D(z)$ is of the form

$$D(z) = 1+\bar{c}_2 z^2+\bar{c}_3 z^3+\ldots, \tag{5.13}$$

where $\bar{c}_2$, $\bar{c}_3$,... are real numbers.

Truncated left-hand cut

We now assume that $\rho(s) = 0$ outside the segment $s_2 \leqslant s \leqslant s_1$ (where $s_1 \leqslant 0$). This segment can also be written $p_1 \leqslant z \leqslant p_2$ where

$$p_i = |1-\tfrac{1}{4}s_i|^{\frac{1}{2}} \quad (i = 1, 2).$$

Clearly $p_i \geqslant 1$.

Now eqns (5.4) and (5.9) show that

$$|N(z)| = O(|z|^{-2}), \quad \text{for } |z| \to \infty$$

(in the whole z-plane). By eqn (5.5) and eqns (I.7) and (I.8) of Appendix I,

$$|D(\nu)| = O(\ln|\nu|), \quad \text{for } |\nu| \to \infty,$$

on the physical sheet. Equation (5.11) shows that

$$|D(z)| = O(\ln|z|), \quad \text{for } |z| \to \infty$$

(in the whole z-plane).

By eqn (5.13) the function

$$\frac{D(z)-1}{z^2}$$

is regular at $z = 0$. As $|z| \to \infty$ it is $O(\ln|z|/|z|^2)$, so we have the dispersion relation

$$\frac{D(z)-1}{z^2} = \frac{1}{\pi}\int\limits_{-\infty}^{-1} \frac{\operatorname{im} D(z'+)}{z'^2(z'-z)}\,dz'. \tag{5.14}$$

For $-\infty \leqslant z \leqslant -1$, eqn (5.11) gives

$$\operatorname{im} D(z+) = 2|z|^3 \operatorname{im}(N(-z-)/\omega(z+)).$$

Using Fig. 5.1 we have

$$\operatorname{im} D(z+) = -\frac{2|z|^3}{|\omega|}\operatorname{re} N(z). \tag{5.15}$$

Hence eqn (5.14) becomes

$$\frac{D(z)-1}{z^2} = \frac{2}{\pi}\int\limits_1^\infty \frac{z'|z'^2-1|^{-\frac{1}{2}}\,\mathrm{re}\,N(z')}{z'+z}\,\mathrm{d}z'. \tag{5.16}$$

We can write eqn (5.4) in the form

$$N(z) = -\frac{2}{\pi}\int\limits_{p_1}^{p_2} \frac{z'\rho(z')D(z')}{z'^2-z^2}\,\mathrm{d}z'. \tag{5.17}$$

Inserting eqn (5.17) in eqn (5.16) gives

$$\frac{D(z)-1}{z^2} = -\frac{4}{\pi^2}\int\limits_{p_1}^{p_2} K(z,z')z'\rho(z')D(z')\,\mathrm{d}z', \tag{5.18}$$

where
$$K(z,z') = -P\int\limits_1^\infty \frac{t|t^2-1|^{-\frac{1}{2}}\,\mathrm{d}t}{(z+t)(t^2-z'^2)}. \tag{5.19}$$

In eqn (5.19) the principal value symbol refers to the integration past $t = z'$. It follows that

$$K(z,z') = \frac{1}{z+z'}\int\limits_1^\infty \frac{t|t^2-1|^{-\frac{1}{2}}\,\mathrm{d}t}{(z+t)(t+z')} - \frac{1}{z+z'}P\int\limits_1^\infty \frac{t|t^2-1|^{-\frac{1}{2}}}{t^2-z'^2}\,\mathrm{d}t.$$

The principal value integral in this equation vanishes (for $1 \leqslant z' \leqslant \infty$). Thus we have the integral representation

$$K(z,z') = \frac{1}{z+z'}\int\limits_1^\infty \frac{t|t^2-1|^{-\frac{1}{2}}\,\mathrm{d}t}{(z+t)(t+z')} \quad (1 \leqslant z' \leqslant \infty). \tag{5.20}$$

This shows that
$$K(z,z') > 0; \quad z,z' \in [1,\infty]. \tag{5.21}$$

On writing
$$\frac{t}{(z+t)(t+z')} = \frac{1}{z-z'}\left(\frac{z}{t+z} - \frac{z'}{t+z'}\right),$$

eqn (5.20) becomes

$$K(z,z') = \frac{g(z)-g(z')}{z^2-z'^2} \quad (1 \leqslant z' \leqslant \infty), \tag{5.22}$$

where
$$g(z) = z\int\limits_1^\infty \frac{|t^2-1|^{-\frac{1}{2}}}{t+z}\,\mathrm{d}t \tag{5.23}$$

$$= -i(z/\omega)\ln(z+i\omega). \tag{5.24}$$

The ln term is defined to be positive when its argument is real and exceeds unity. Equations (5.18) and (5.22) have been found by Chew and Mandelstam [25].

Equation (5.18) is an integral equation for $D(z)$ on $p_1 \leqslant z \leqslant p_2$. On solving that equation, the value of $D(z)$ for any z is also found from eqn (5.18).

5.3. The function $g(z)$

We shall now study some properties of the function $g(z)$. It follows from the integral representation, eqn (5.23), that

$$g^*(z) = g(z^*),$$

and that $g(z)$ is regular in the z-plane cut along $-\infty \leqslant z \leqslant -1$. Also

$$\operatorname{im} g(z+) = -\frac{\pi z}{|z^2-1|^{\frac{1}{2}}}, \quad \text{for } -\infty \leqslant z \leqslant -1,$$

so $g(z)$ has a square root branch point at $z = -1$. Since

$$\operatorname{im} g(z) = \operatorname{im} z \,.\int\limits_1^\infty \frac{t|t^2-1|^{-\frac{1}{2}}}{|t+z|^2}\,dt,$$

it follows that
$$\operatorname{im} g(z) > 0, \quad \text{for } \operatorname{im} z > 0. \tag{5.25}$$

To study the explicit form of $g(z)$ (eqn (5.24)) it is convenient to make a conformal mapping of the z-plane on to the plane $\phi = u+iv$. The transformation

$$z = \sin\phi \tag{5.26}$$

maps the z-plane cut along $-\infty \leqslant z \leqslant -1$ and $1 \leqslant z \leqslant \infty$ on to the strip $-\frac{1}{2}\pi \leqslant u \leqslant \frac{1}{2}\pi$, $-\infty \leqslant v \leqslant \infty$ (Fig. 5.2). Equations (5.8) and (5.26) give

$$\omega = \cos\phi, \tag{5.27}$$

and it is easy to see that this agrees with the sign convention shown in Fig. 5.1.

Equation (5.24) can be written

$$g(z) = (\tfrac{1}{2}\pi-\phi)\tan\phi. \tag{5.28}$$

This gives

$$\operatorname{re} g(z) = \frac{(\frac{1}{2}\pi-u)\sin(2u)+v\sinh(2v)}{\cosh(2v)+\cos(2u)}, \tag{5.29}$$

$$\operatorname{im} g(z) = \frac{(\frac{1}{2}\pi-u)\sinh(2v)-v\sin(2u)}{\cosh(2v)+\cos(2u)}. \tag{5.30}$$

Values of $g(z)$ on the real and imaginary axes

Simple algebra gives the following results:

For $-\infty \leqslant z \leqslant -1$,

$$\operatorname{im} g(z+) = \frac{\pi|z|}{|z^2-1|^{\frac{1}{2}}}, \tag{5.31}$$

$$\operatorname{re} g(z) = \frac{|z|}{|z^2-1|^{\frac{1}{2}}}\ln(|z|+|z^2-1|^{\frac{1}{2}}), \tag{5.32}$$

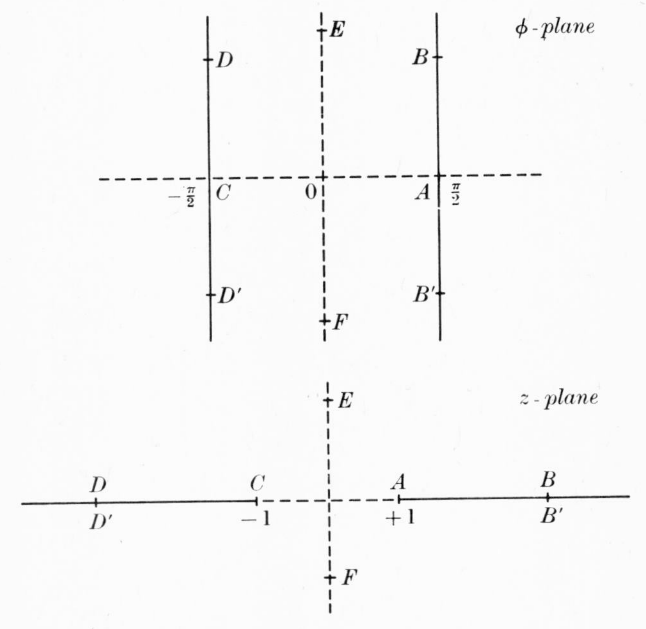

FIG. 5.2. The mapping $z = \sin\phi$ of the cut z-plane on to the strip $-\frac{1}{2}\pi \leqslant u \leqslant \frac{1}{2}\pi$, $-\infty \leqslant v \leqslant \infty$ of the ϕ-plane. Corresponding points are indicated.

also $$\operatorname{re} g(z) \to 1, \quad \text{as } z \to -1 \text{ (from below).}$$

For $-1 \leqslant z \leqslant +1$,

$$g(z) = \frac{z}{|1-z^2|^{\frac{1}{2}}}\{\tfrac{1}{2}\pi - \arcsin z\}; \tag{5.33}$$

also $$g(1) = 1, \qquad g(0) = 0,$$

$$g(z) \to -\infty, \quad \text{as } z \to -1 \quad \text{(from above).}$$

For $1 < z \leqslant \infty$, $$g(z) = \frac{z}{|z^2-1|^{\frac{1}{2}}}\ln(z+|z^2-1|^{\frac{1}{2}}), \tag{5.34}$$

and $$g(z) \to \ln(2z), \quad \text{as } z \to +\infty;$$

also $$\operatorname{re} g(-z) = g(z). \tag{5.35}$$

Equation (5.23) gives

$$\frac{\mathrm{d}g(z)}{\mathrm{d}z} = \int\limits_1^\infty \frac{t|t^2-1|^{-\frac{1}{2}}\,\mathrm{d}t}{(z+t)^2},$$

so $$\mathrm{d}g(z)/\mathrm{d}z > 0, \quad \text{for } -1 < z < \infty; \tag{5.36}$$

the values of $\operatorname{re} g(z)$ on the real axis are shown in Fig. 5.3.

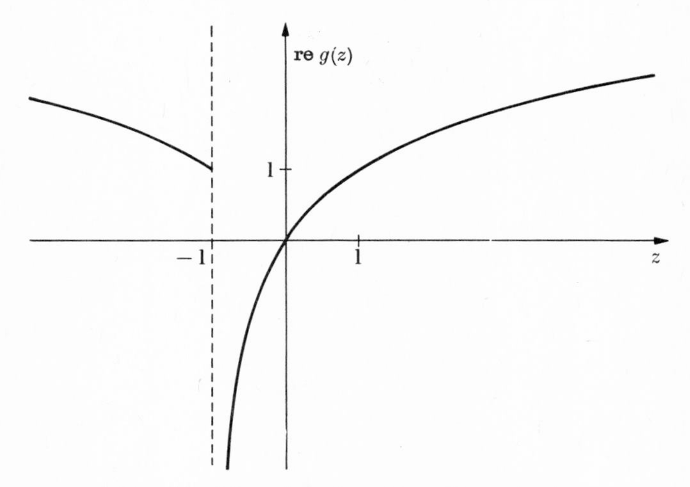

FIG. 5.3. The value of re $g(z)$ on the real axis.

On $\operatorname{re} z = 0$.

Here $u = 0$ and $z = iy = i \sinh v$, and we have

$$\operatorname{re} g(iy) = \frac{|y|}{|y^2+1|^{\frac{1}{2}}} \ln(|y|+|y^2+1|^{\frac{1}{2}}), \tag{5.37}$$

$$\operatorname{im} g(iy) = \tfrac{1}{2}\pi \frac{y}{|y^2+1|^{\frac{1}{2}}}. \tag{5.38}$$

For $|z| < 1$

we can expand eqn (5.23) in a Taylor series,

$$\begin{aligned} g(z) &= -\sum_{n=1}^{\infty} (-z)^n \int_1^{\infty} \frac{\mathrm{d}t}{t^n(t^2-1)^{\frac{1}{2}}} \\ &= -\sum_{n=1}^{\infty} (-z)^n \int_0^{\frac{1}{2}\pi} \cos^{(n-1)}\psi \, \mathrm{d}\psi \\ &= -\frac{1}{2} \sum_{n=1}^{\infty} (-z)^n \frac{\Gamma(\frac{1}{2}n)\Gamma(\frac{1}{2})}{\Gamma(\frac{1}{2}(n+1))} \\ &= \tfrac{1}{2}\pi z - z^2 + \tfrac{1}{4}\pi z^3 - \tfrac{2}{3}z^4 + \dots . \end{aligned} \tag{5.39}$$

5.4. The locus $\operatorname{im} D(z, \lambda) = 0$

As in Chapter 4 we make it possible to vary the strength of the left-hand cut discontinuity by writing

$$\rho(s) = -\lambda\bar{\rho}(z),$$

where λ is a positive parameter. The class of truncated attractive left-hand cuts which we consider have

$$\bar{\rho}(z) \geqslant 0, \quad \text{for } 1 \leqslant p_1 \leqslant z \leqslant p_2,$$

and $\bar{\rho}(z) = 0$ elsewhere.

Equation (5.18) becomes

$$D(z,\lambda) = 1+\frac{4\lambda z^2}{\pi^2}\int\limits_{p_1}^{p_2} K(z,z')z'\bar{\rho}(z')D(z',\lambda)\,\mathrm{d}z'. \tag{5.40}$$

Bearing in mind eqn (5.21), the method of Appendix III can be used to show that there is a range $0 < \lambda < \lambda_1$ for which $D(z,\lambda) > 1$ in $p_1 \leqslant z \leqslant p_2$. For such values of z, $D(z,\lambda)$ is a monotonic increasing function of λ such that $D(z,\lambda) \to +\infty$ as $\lambda \to \lambda_1$. Thus at $\lambda = \lambda_1$, a bound state (or resonance) appears at the physical threshold $z = 0$.

We are concerned with the range $0 < \lambda < \lambda_1$. There eqn (5.40) is of the form

$$D(z,\lambda) = 1+z^2\int\limits_{p_1}^{p_2} K(z,z')\sigma(z')\,\mathrm{d}z', \tag{5.41}$$

where $$\sigma(z') = \frac{4\lambda}{\pi^2}z'\bar{\rho}(z')D(z',\lambda), \quad \text{for } p_1 \leqslant z' \leqslant p_2,$$

so $$\sigma(z') \geqslant 0. \tag{5.42}$$

The locus $\operatorname{im} D(z,\lambda) = 0$ is given by

$$\int\limits_{p_1}^{p_2} I(z,z')\sigma(z')\,\mathrm{d}z' = 0, \tag{5.43}$$

where
$$\begin{aligned} I(z,z') &= \operatorname{im}(z^2K(z,z')) \\ &= \operatorname{im}\left(z^2\frac{g(z)-g(z')}{z^2-z'^2}\right), \end{aligned} \tag{5.44}$$

and $p_1 \leqslant z' \leqslant p_2$.

On $-1 \leqslant z \leqslant \infty$, $g(z)$ is real, so this line is one part of the locus $\operatorname{im} D(z,\lambda) = 0$. In the s-plane this gives the lines $-\infty \leqslant s \leqslant s_0$ on the physical sheet and $0 \leqslant s \leqslant s_0$ on the second sheet.

By eqns (5.31), (5.42), and (5.43) it is clear that the locus $\operatorname{im} D(z,\lambda) = 0$ can only meet the line $-\infty \leqslant z \leqslant -1$ in isolated points which must lie in $-p_2 \leqslant z \leqslant -p_1$.

In order to study the location of $\operatorname{im} D(z,\lambda) = 0$ off the real axis we shall begin with a single driving pole.

Single driving pole

We put $\bar{\rho}(z') \propto \delta(z'-p)$ where $p \geqslant 1$. Then eqn (5.43) becomes

$$I(z,p) = 0. \tag{5.45}$$

Since this equation does not depend on λ it must contain the locus of the zeros $D(z,\lambda) = 0$ as λ varies (this is a special property of the single pole case).

One branch of the locus given by eqn (5.45) is the line $-1 \leqslant z \leqslant \infty$. The other branch lies off the real axis. We shall call this the *second branch*, and now examine its properties. Equation (5.44) and the representation in eqn (5.20) give

$$I(z,p) = \frac{\operatorname{im} z}{|z+p|^2} \int\limits_1^\infty \frac{\{|z|^2(t+p)+2pt\operatorname{re} z\}t(t^2-1)^{-\frac{1}{2}}}{|z+t|^2(t+p)}\,dt, \qquad (5.46)$$

so the second branch of $I(z,p) = 0$ is given by

$$|z|^2p\langle t^{-1}\rangle + |z|^2 + 2p\operatorname{re} z = 0, \qquad (5.47)$$

where $$\langle t^{-1}\rangle = \int\limits_1^\infty t^{-1}\sigma(t;z,p)\,dt \Big/ \int\limits_1^\infty \sigma(t;z,p)\,dt$$

and $$\sigma(t;z,p) = \frac{t^2(t^2-1)^{-\frac{1}{2}}}{|z+t|^2(t+p)}.$$

Since $$0 \leqslant \langle t^{-1}\rangle \leqslant 1,$$

eqn (5.47) shows that the second branch of $I(z,p) = 0$ must lie between the circles

$$|z+p| = p \quad \text{and} \quad \left|z+\frac{p}{p+1}\right| = \frac{p}{p+1}. \qquad (5.48)$$

The region where $I(z,p) = 0$ can be further restricted. Equation (5.45) is written as

$$\operatorname{im}\Psi(z,p) = 0, \qquad (5.49)$$

where $$\Psi(z,p) = (|z|^4 - p^2z^2)(g(z)-g(p)). \qquad (5.50)$$

This gives† $$|z|^2 - p^2\cos(2\psi) = p^2\frac{\sin(2\psi)}{\operatorname{im} g(z)}\{\operatorname{re} g(z) - g(p)\}. \qquad (5.51)$$

By eqn (5.25), $$\sin(2\psi)/\operatorname{im} g(z) < 0$$

in $\operatorname{re} z < 0$.

The level curve

$$\operatorname{re} g(z) = g(p), \quad \text{for } p \geqslant 1, \qquad (5.52)$$

makes a circuit from the negative axis‡ at $z = -p$, passing through $z = iy_0$, $z = +p$, $z = -iy_0$ and ending at $z = -p$ again. By eqns (5.34) and (5.37), y_0 is given by

$$\frac{y_0}{(y_0^2+1)^{\frac{1}{2}}}\ln\{y_0+(y_0^2+1)^{\frac{1}{2}}\} = \frac{p}{(p^2-1)^{\frac{1}{2}}}\ln\{p+(p^2-1)^{\frac{1}{2}}\}. \qquad (5.53)$$

As we shall see in § 5.7 below, $y_0 > p$.

† As usual $z = |z|\exp(i\psi)$.

‡ The tangents at $z = -p$ lie in the region $\operatorname{re} z < -p$.

From eqn (5.29) it is clear that $\operatorname{re} g(z) > g(p)$ when z is outside the region bounded by eqn (5.52). Thus the second branch of $I(z,p) = 0$ must lie between the level curve, eqn (5.52), and the lemniscate

$$|z|^2 = p^2 \cos(2\psi). \tag{5.54}$$

In Fig. 5.4 we show an example of the region in which $I(z,p) = 0$ must lie.

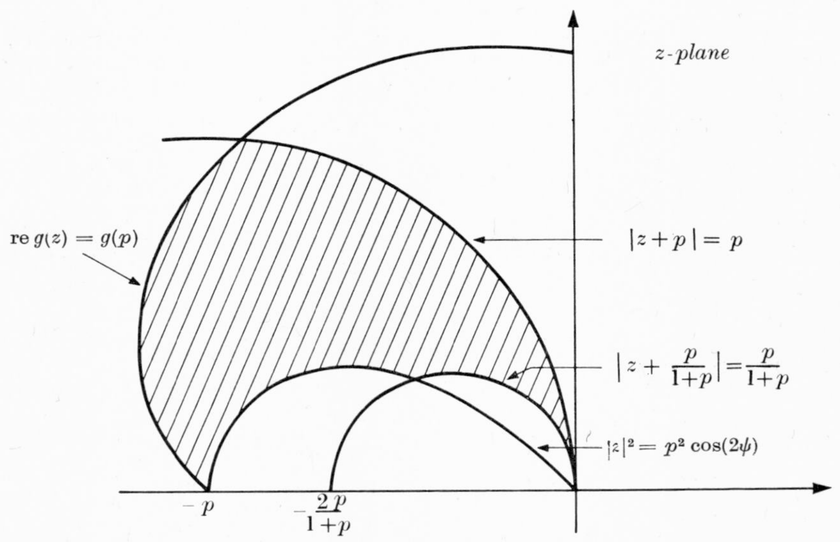

FIG. 5.4. The shaded region, given by eqns (5.48), (5.52), and (5.54), shows where the second branch of $I(z, p) = 0$ must lie for $\operatorname{im} z > 0$.

The curve $I(z,p) = 0$ passes through $z = 0$, and eqns (5.48) show that at $z = 0$ it touches the axis $\operatorname{re} z = 0$. Using eqns (5.39) and (5.44) we find that for small $|z|$ the curve has the form

$$(\operatorname{im} z)^2 + \frac{4}{\pi} g(p) \operatorname{re} z = 0. \tag{5.55}$$

Thus its radius of curvature at $z = 0$ is $R = (2/\pi)g(p)$. For $p \to 1$, $R \to (2/\pi)$, while for large p, $R \simeq (2/\pi)\ln(2p)$.

The family $I(z,p) = 0$

From eqn (5.46) it follows that on $I(z,p) = 0$,

$$\frac{\partial}{\partial p} I(z,p) = \frac{2 \operatorname{im} z \operatorname{re} z}{|z+p|^2} \int_1^\infty \frac{t^3(t^2-1)^{-\frac{1}{2}}}{|z+t|^2(t+p)^2}\, dt;$$

therefore

$$\operatorname{im} z \, . \, \frac{\partial I(z,p)}{\partial p} < 0$$

on the second branch of $I(z,p) = 0$. It follows that if $p' > p$, the second branch of $I(z,p') = 0$ lies wholly outside the second branch of $I(z,p) = 0$ (except as $z \to 0$).

Typical second branches of $I(z,p) = 0$ are shown in Fig. 5.5. At $z = -p$ they are orthogonal to the real axis, except for $p = 1$ in which case the angle is $\frac{1}{3}\pi$.

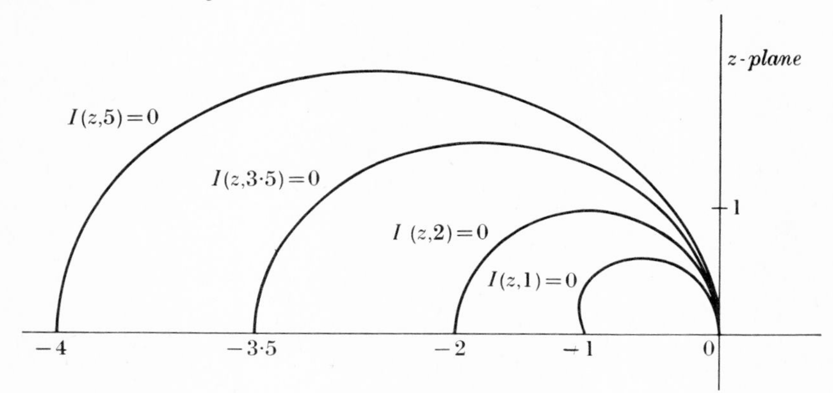

FIG. 5.5. Typical second branches of $I(z, p) = 0$.

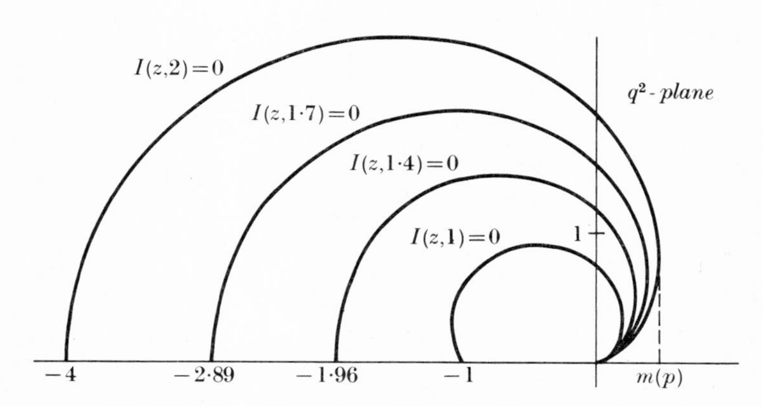

FIG. 5.6. Second branches of $I(z, p) = 0$ in the q^2-plane.

Fig. 5.6 shows the second branch of $I(z,p) = 0$ plotted in the q^2-plane. Let $m(p)$ be the maximum of $\text{re}(q^2)$ on $I(z,p) = 0$. It follows from the discussion that $m(p)$ is a monotonic increasing function of p. For large p, $m(p) \simeq 0{\cdot}042p^2$. Values of $m(p)$ are shown in Fig. 5.7 (a).

Bounds on $\text{im}\, D(z, \lambda) = 0$

It follows from eqn (5.43) that for $0 < \lambda < \lambda_1$, and $\text{im}\, z \neq 0$, the locus

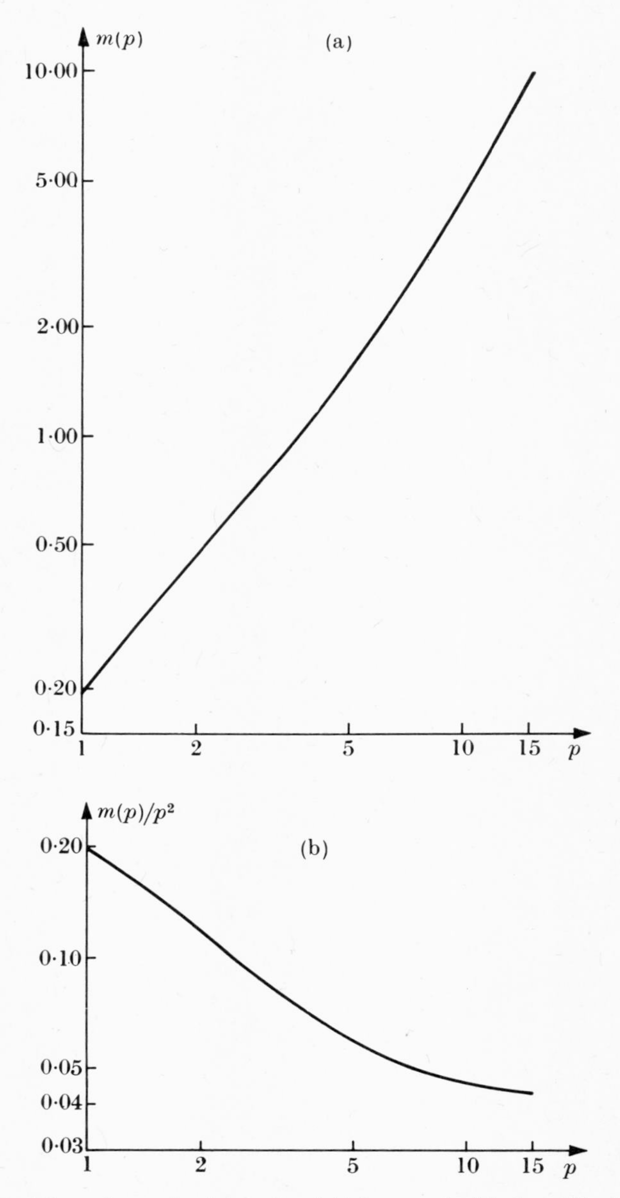

FIG. 5.7. The values of $m(p)$ and $m(p)/p^2$ on a double logarithmic scale.

$\text{im}\, D(z, \lambda) = 0$ must lie in the region bounded by the second branches of

$$I(z, p_1) = 0 \quad \text{and} \quad I(z, p_2) = 0.$$

This is obviously the optimum result which can be obtained without further specifying the discontinuity $\bar{\rho}(z)$ for $p_1 \leqslant z \leqslant p_2$.

The maximum value of $\text{re}\, \tilde{s}$, where $\tilde{s}$ is the position of the second sheet pole, will obey

$$(\text{re}\, \tilde{s})_{\max} \leqslant s_0 + 4m(p_2). \tag{5.56}$$

Comparison with the e.m.h. case

It is interesting to compare these results with those in Chapter 4. The relation between q^2 and s is the same in both cases but z is different. Consider a single driving pole at $s_0 - p^2$ in both cases. We have to take $p \geqslant 2\mu$ in order to have a corresponding π–π case.

As λ varies the second sheet pole $\tilde{s}$ will leave the real axis at $s_0 - 4p^2$ in the e.m.h. case (Fig. 4.3). In the π–π case it leaves the real axis at $s_0 - p^2$. The maximum values of $\text{re}\, \tilde{s}$ are

$$\text{e.m.h.:} \quad s_0 + \tfrac{1}{2}p^2;$$

$$\pi\text{–}\pi\text{:} \quad s_0 + 4m(\tfrac{1}{2}p).$$

These two expressions may be compared by writing the second as

$$\pi\text{–}\pi\text{:} \quad s_0 + p^2\{m(\tfrac{1}{2}p)/(\tfrac{1}{2}p)^2\}.$$

The function $m(p')/p'^2$ is plotted in Fig. 5.7 (b); its value is 0·197 for $p' = 1$, and it falls quickly to reach 0·042 for large p'.

5.5. Multiple loops

We saw in § 5.4 that the locus $\text{im}\, D(z, \lambda) = 0$ can only meet the line $-\infty \leqslant z \leqslant -1$ in isolated points $\bar{z}_i$ on $[-p_2, -p_1]$. Equation (5.15) shows that these are the points where $\text{re}\, N(z)$ vanishes. Equation (5.17) shows that $N(z) < 0$ for $-\infty < z < -p_2$, and $N(z) > 0$ for $-p_1 < z < 0$, so there must be an odd number of points $\bar{z}_i$. Therefore an odd number of branches of $\text{im}\, D(z, \lambda) = 0$ leave $[-p_2, -p_1]$.

Two branches of $\text{im}\, D(z, \lambda) = 0$ cannot meet and have a common tangent at a point z where $D(z, \lambda)$ is regular. Therefore only one branch of $\text{im}\, D(z, \lambda) = 0$ crosses the real axis at $z = 0$. Only one of the branches which leave the interval $[-p_2, -p_1]$ returns to the real axis at $z = 0$; the others must return to the interval $[-p_2, -p_1]$.

A possible behaviour of the branches of $\text{im}\, D(z, \lambda) = 0$ is shown in Fig. 5.8. This is more complicated than corresponding figures in the e.m.h. case (for example, Figs. 4.6 (a) and 4.6 (b)). We saw in the e.m.h. case (§ 4.6) that, for fixed λ in $0 < \lambda < \lambda_1$, there is at most one pair of

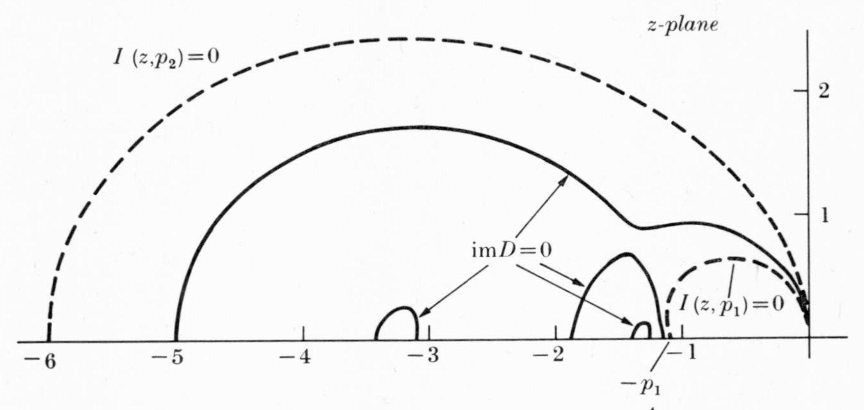

FIG. 5.8. A possible behaviour of $\operatorname{im} D = 0$. Here $\bar{\rho} = \sum_{i=1}^{4} a_i\, \delta(z-p_i)$.

complex conjugate zeros $D(z, \lambda) = 0$. In the π–π case this is not true, and we can give fairly simple examples which have *several pairs of complex conjugate zeros* $D(z, \lambda) = 0$.

We saw that only one branch of $\operatorname{im} D(z, \lambda) = 0$ can pass through $z = 0$ from $\operatorname{im} z > 0$. It is therefore a matter of some practical importance to determine how close to $\operatorname{re} z = 0$ the other branches, or loops, of $\operatorname{im} D(z, \lambda) = 0$ can approach.

It is convenient to study this question in the ϕ-plane. The example shown in Fig. 5.8 then appears as in Fig. 5.9; we have used the notation

$$v(p) = \operatorname{arc\,cosh} p.$$

A point where the tangent to a branch of $\operatorname{im} D(z, \lambda) = 0$ is parallel to the v-axis will be called a *turning point*.

Consider the attractive interaction given by

$$\bar{\rho}(z) = \alpha\delta(z-p_1)+\beta\delta(z-p_2), \tag{5.57}$$

where α and β are positive numbers. Then $\operatorname{im} D(z, \lambda) = 0$ is

$$\alpha' I(z, p_1)+\beta' I(z, p_2) = 0 \tag{5.58}$$

where
$$\frac{\alpha'}{\beta'} = \frac{\alpha}{\beta}\frac{p_1}{p_2}\frac{D(p_1, \lambda)}{D(p_2, \lambda)}.$$

At a turning point

$$\alpha' \frac{\partial}{\partial v} I(z, p_1)+\beta' \frac{\partial}{\partial v} I(z, p_2) = 0. \tag{5.59}$$

Eliminating α' and β' from eqns (5.58) and (5.59) gives

$$J(z; p_1, p_2) = 0, \tag{5.60}$$

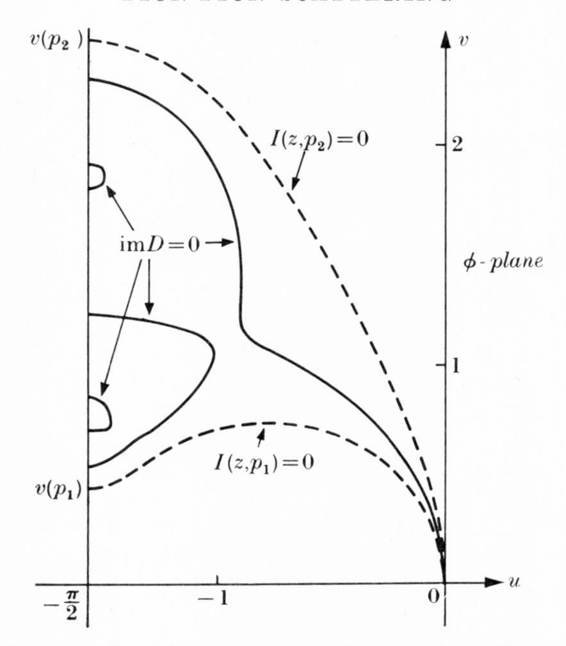

FIG. 5.9. Branches of im $D = 0$ shown in the ϕ-plane.

where

$$J(z;z',z'') = \begin{vmatrix} I(z,z') & I(z,z'') \\ \dfrac{\partial}{\partial v} I(z,z') & \dfrac{\partial}{\partial v} I(z,z'') \end{vmatrix}. \tag{5.61}$$

We are only interested in the portion of $J(z;p_1,p_2) = 0$ which lies in the region bounded by the line $u = -\frac{1}{2}\pi$ and the second branches of $I(z,p_1) = 0$, $I(z,p_2) = 0$, i.e. the portion where $J(z;p_1,p_2) = 0$ and $I(z,p_1).I(z,p_2) < 0$. An example is shown in Fig. 5.10. We can now choose α/β such that a branch of im $D(z,\lambda) = 0$ has its turning point at any position on this portion of $J(z;p_1,p_2) = 0$.

Let $\bar{u}(p_1, p_2)$ be the maximum value of u on this portion of $J(z;p_1,p_2) = 0$. For the interaction in eqn (5.57) there can only be one branch of im $D(z,\lambda) = 0$ in the region between re $z = 0$ and the hyperbola

$$\left(\frac{\text{re}\, z}{\sin \bar{u}}\right)^2 - \left(\frac{\text{im}\, z}{\cos \bar{u}}\right)^2 = 1 \tag{5.62}$$

in the z-plane (see Fig. 5.11).

For example if $\bar{u} = -\frac{1}{4}\pi$, there can only be one pair of complex con jugate zeros $D(z,\lambda) = 0$ in the region re $q^2 > -\frac{1}{2}$, i.e. in the region re $s > 2\mu^2$.

Let $E(p_1,p_2)$ be the region where the turning points can lie for the

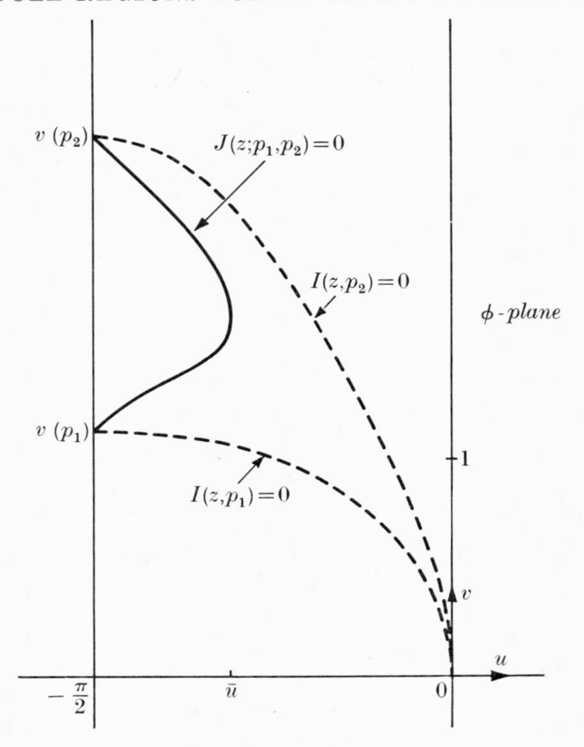

FIG. 5.10. The relevant branch of $J(z\,; p_1, p_2) = 0$ on which lie the turning points of the curves $\operatorname{im} D(z, \lambda) = 0$ arising from $\bar{\rho}(z)$ of eqn (5.57).

general $\bar{\rho}(z)$ $(p_1 \leqslant z \leqslant p_2)$. In Appendix IV we show that $E(p_1, p_2)$ is the region covered by the family of curves

$$J(z; z', p_2) = 0, \quad I(z, z') \,.\, I(z, p_2) < 0,$$

for $p_1 \leqslant z' \leqslant p_2$, and the family of curves

$$J(z; p_1, z') = 0, \quad I(z, p_1) \,.\, I(z, z') < 0,$$

for $p_1 \leqslant z' \leqslant p_2$. We also show that the maximum value of u on $E(p_1, p_2)$ is $\bar{u}(p_1, p_2)$. Then again only one branch of $\operatorname{im} D(z, \lambda) = 0$ can lie in the region $\operatorname{im} z > 0$ between $\operatorname{re} z = 0$ and the hyperbola in eqn (5.62).

In Fig. 5.12 we show $\bar{u}(p_1, p_2)$ for various values of p_1, p_2. Notice for example that $\bar{u} = -\frac{1}{4}\pi$ for $p_1 = 1$, $p_2 = 6{\cdot}3$. For fixed p_1, $\bar{u}$ increases as p_2 increases, while for fixed p_2, $\bar{u}$ decreases as p_1 increases.

These results are the best possible without further specifying $\bar{\rho}(z)$. We can give left-hand cut functions $\bar{\rho}(z)$ such that for some λ $(0 < \lambda < \lambda_1)$ there are several pairs of complex zeros $D(z, \lambda) = 0$, one pair being at a turning point of a loop of $\operatorname{im} D(z, \lambda) = 0$, and another pair being on the branch which goes through $z = 0$.

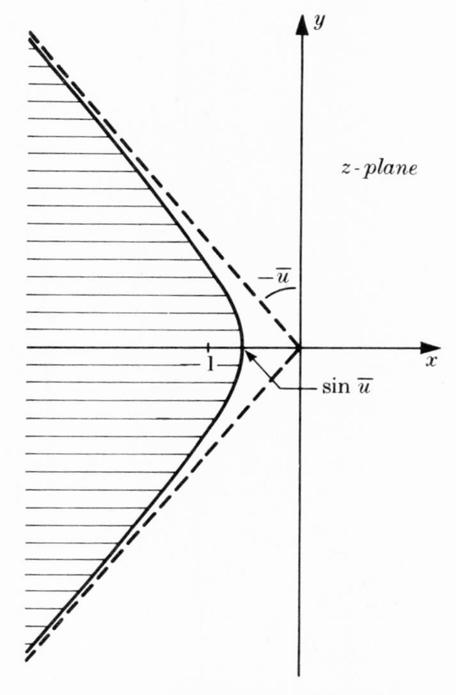

FIG. 5.11. Only one second branch of im $D(z, \lambda) = 0$ can lie off the real axis to the right of the hyperbola.

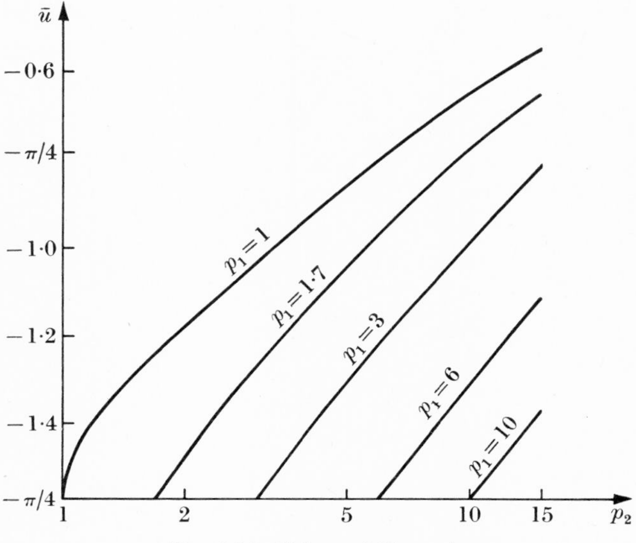

FIG. 5.12. Values of $\bar{u}(p_1, p_2)$.

5.6. Motion of the second sheet poles

$\operatorname{re} D(z,\lambda)$ increases monotonically as z moves along a branch of $\operatorname{im} D(z,\lambda) = 0$ such that $\operatorname{im} D(z,\lambda) > 0$ to the left. Thus $\operatorname{re} D(z,\lambda)$ decreases on $(-1,0)$ and increases on $(0,\infty)$, and since $D(0) = 1$, there cannot be any zeros of $D(z,\lambda)$ on $-1 < z < \infty$ (for $0 \leqslant \lambda < \lambda_1$).

We saw in § 4.9 in the e.m.h. case that as λ increases from 0, a zero of $D(z,\lambda)$ moves in from $z = -\infty$ along the real axis. In the π–π case it is very different as the following proof for Theorem 11 shows.

THEOREM 11. *Assume that* $|z^2-1|^{-\frac{1}{2}}\bar{\rho}(z)$ *is continuous in* $[p_1, p_2]$ *and let*

$$M = \max\{2z^3|z^2-1|^{-\frac{1}{2}}\bar{\rho}(z)\}.$$

Then $\lambda_1 > M^{-1}$, *and* $D(z,\lambda)$ *has no zeros for* $0 < \lambda < M^{-1}$.

Let $\bar{z}$ be any point for which $\operatorname{im} D(\bar{z},\lambda) = 0$ and $\operatorname{im} \bar{z} \neq 0$. We can always move from $\bar{z}$ to some point $\bar{\bar{z}}$ on the interval $[-p_2, -p_1]$ following some branch of $\operatorname{im} D(z,\lambda) = 0$ in the direction of decreasing $\operatorname{re} D(z,\lambda)$. Therefore

$$\operatorname{re} D(\bar{z},\lambda) > \operatorname{re} D(\bar{\bar{z}},\lambda).$$

By eqns (5.11) and (5.17),

$$\operatorname{re} D(-z,\lambda) = D(z,\lambda)\left\{1 - \frac{2\lambda z^3}{|z^2-1|^{\frac{1}{2}}}\bar{\rho}(z)\right\}, \quad \text{for } p_1 \leqslant z \leqslant p_2. \tag{5.63}$$

If λ is close to, but less than, λ_1 there is a zero of $D(z,\lambda)$ near $z = 0$. So $\operatorname{re} D(\bar{\bar{z}},\lambda) < 0$ for some $\bar{\bar{z}}$ in $[-p_2, -p_1]$. Since $D(z,\lambda) > 1$ on $[p_1, p_2]$ and $D(z,\lambda) \to +\infty$ as $\lambda \to \lambda_1$, it follows from eqn (5.63) that $\lambda_1 > M^{-1}$.

For $0 < \lambda < M^{-1}$ eqn (5.63) now shows that $\operatorname{re} D(z,\lambda) > 0$ on $[-p_2, -p_1]$. This proves the theorem.

As λ increases above M^{-1}, $\operatorname{re} D(z,\lambda)$ becomes negative on increasing sub-intervals of $[-p_2, -p_1]$. When $\operatorname{re} D(z,\lambda)$ goes through zero where a branch of $\operatorname{im} D(z,\lambda) = 0$ meets the real axis, a zero of $D(z,\lambda)$ will leave, or return to, the real axis according to whether $\operatorname{re} D(z,\lambda)$ increases or decreases as z moves into the complex plane along the branch.

Equation (5.55) shows that near $z = 0$ the complex branch of $\operatorname{im} D(z,y) = 0$ lies between the parabolas

$$(\operatorname{im} z)^2 = -a_1 \operatorname{re} z \quad \text{and} \quad (\operatorname{im} z)^2 = -a_2 \operatorname{re} z,$$

where

$$a_i = \frac{4}{\pi} g(p_i) \quad (i = 1, 2).$$

When λ is close to λ_1 (and $\lambda < \lambda_1$) the zeros of $D(z,\lambda)$ are at $z = \pm iy + O(y^2)$

where by eqns (5.22), (5.38), and (5.40)

$$\frac{1}{y^2} = \frac{4\lambda_1}{\pi^2} \int_{p_1}^{p_2} dz \, \frac{g(z)}{z} \, \bar{\rho}(z) D(z, \lambda).$$

Since, as $\lambda \to \lambda_1$,

$$D(z, \lambda) \to \frac{c(z)}{\lambda_1 - \lambda}, \quad \text{for } p_1 \leqslant z \leqslant p_2,$$

where $c(z)$ is positive and bounded (see Theorem C of Appendix III), the position of the pole $\tilde{s}$ obeys

$$|\tilde{s} - s_0| = \pi^2 \left(1 - \frac{\lambda}{\lambda_1}\right) \Big/ \int_{p_1}^{p_2} dz \, \frac{g(z)}{z} \, \bar{\rho}(z) c(z). \tag{5.64}$$

Using the relation

$$\operatorname{im} \tilde{s} = -4 \operatorname{im}(z^2) = -8y \operatorname{re} z,$$

we find that as $\lambda \to \lambda_1$,

$$\frac{1}{a_2} \leqslant \frac{\operatorname{im} \tilde{s}}{|\tilde{s} - s_0|^{\frac{3}{2}}} \leqslant \frac{1}{a_1}. \tag{5.65}$$

Since $g(p)$ (eqn (5.34)) is a slowly varying function of p, eqn (5.65) can provide useful information on $\operatorname{im} \tilde{s}$.

5.7. The position of the resonance

The phase shift

The phase shift $\alpha(s)$ is given by

$$q^3 \cot \alpha(s) = \tfrac{1}{2} s^{\frac{1}{2}} \operatorname{re} D(s)/N(s). \tag{5.66}$$

On the physical cut $s_0 \leqslant s \leqslant \infty$, the point $s+$ corresponds to $z = -iy$ $(0 \leqslant y \leqslant \infty)$. Remembering that $\rho(s) = -\lambda \bar{\rho}(z)$, it follows from eqn (5.17) that, if $0 < \lambda < \lambda_1$,

$$N(s) > 0, \quad \text{for } s_0 \leqslant s < \infty.$$

Now eqn (5.66) shows that

$$\alpha(s) \neq 0 \pmod{\pi}, \quad \text{for } s_0 < s < \infty.$$

The scattering length, defined by the relation $\alpha(s+) = a_{\pi\pi} q^3 + O(q^5)$ for small q, is

$$a_{\pi\pi} = \frac{2}{s_0^{\frac{1}{2}}} N(s_0) > 0,$$

so

$$0 < \alpha(s) < \pi \quad (s_0 < s < \infty). \tag{5.67}$$

We use eqns (5.22) and (5.41) for $z = -iy$, giving

$$D(-iy, \lambda) = 1 + y^2 \int_{p_1}^{p_2} \frac{g(-iy) - g(z')}{y^2 + z'^2} \, \sigma(z') \, dz'. \tag{5.68}$$

Thus
$$\operatorname{re} D(\pm iy, \lambda) = 1 + \int_{p_1}^{p_2} R(y, z')\sigma(z')\, dz', \tag{5.69}$$

where
$$R(y, z') = \frac{y^2}{y^2+z'^2}\{\operatorname{re} g(iy) - g(z')\}, \quad \text{for } p_1 \leqslant z' \leqslant p_2. \tag{5.70}$$

Similarly
$$\operatorname{im} D(-iy, \lambda) = -y^2 \operatorname{im} g(iy) \int_{p_1}^{p_2} \frac{\sigma(z')}{y^2+z'^2}\, dz'. \tag{5.71}$$

Using eqns (5.37) and (5.38) we deduce the limiting behaviour as $y \to \infty$ to be
$$\operatorname{re} D(-iy, \lambda) \to (\ln y) \int_{p_1}^{p_2} \sigma(z')\, dz' + O(1), \tag{5.72}$$
$$\operatorname{im} D(-iy, \lambda) \to -\tfrac{1}{2}\pi \int_{p_1}^{p_2} \sigma(z')\, dz' + O(y^{-1}). \tag{5.73}$$

Equation (5.66) and the relation
$$\tan \alpha(s) = -\frac{\operatorname{im} D(-iy, \lambda)}{\operatorname{re} D(-iy, \lambda)}, \quad \text{for } 0 \leqslant y \leqslant \infty,$$

together with eqns (5.72) and (5.73) show that
$$\alpha(s) \to \frac{\pi}{\ln s} + O((\ln s)^{-2}), \quad \text{as } s \to +\infty. \tag{5.74}$$

Thus $\alpha(s)$ is of the form in eqn (2.87) (§ 2.7) with $\alpha(\infty) = 0$, $H = 1$. Using Theorems 8 and 9 (§ 2.7) this proves that the solution we have considered in the present chapter is the isolated solution of eqn (5.2). Lemmas 5 and 6 show that any other solution must have $\alpha(\infty) = 0$ (mod π) and $H = 0$ or $H = 1$; this result could also be seen by examining the N/D equations we get by introducing extra subtractions into eqns (5.4) and (5.5).

The phase shift behaviour here differs from the e.m.h. case where for the isolated solution $0 < \alpha(s) < \pi$ in $s_0 < s \leqslant \infty$ but $\alpha(\infty) = \tfrac{1}{2}\pi$ (§ 4.10).

Bounds on the resonance position

At a resonance s_R the phase shift passes upwards through $\tfrac{1}{2}\pi$. This requires
$$\operatorname{re} D(s_R) = 0 \quad \text{and} \quad \frac{d}{ds} \operatorname{re} D(s)\bigg|_{s_R} < 0. \tag{5.75}$$

At the end of § 3.2 we showed that a second sheet pole cannot lie near a segment of the physical cut on which $\alpha(s)$ decreases through $\tfrac{1}{2}\pi$. Thus the positions where $\alpha(s)$ passes downwards through $\tfrac{1}{2}\pi$ are of no special physical significance.

We now use eqns (5.69) and (5.70) to study the location of s_R. We always keep λ in the range $0 < \lambda < \lambda_1$. Equation (5.37) shows that $R(y,p)$ $(p \geqslant 1)$ is of the form shown in Fig. 5.13. There are two important parameters $y_0(p)$ and $y_m(p)$ which are defined by

$$R(y_0, p) = 0 \quad (y_0 > 0),$$

$$\frac{\mathrm{d}}{\mathrm{d}y} R(y,p)\bigg|_{y_m} = 0 \quad (y_m > 0).$$

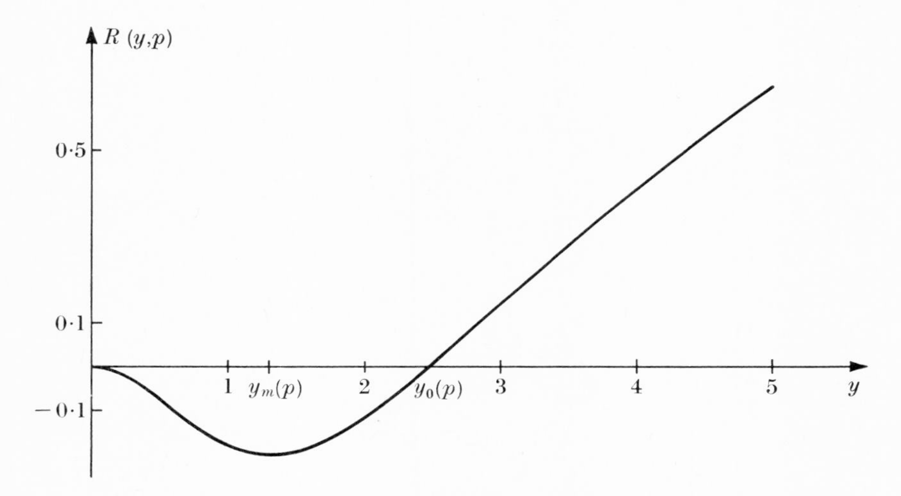

FIG. 5.13. An example of the function $R(y, p)$.

Equation (5.53) gives $y_0(p)$. We find $y_0(p) > p$ for $p \geqslant 1$ and $y_0(1) = 1{\cdot}51$. For large p

$$y_0(p) = p + O\left(\frac{\ln p}{p}\right).$$

Fig. 5.14 shows the numerical values.

For large p

$$y_m(p) = \zeta p + O(p^{-1}),$$

where

$$\ln \zeta + \tfrac{1}{2}(1+\zeta^2) = 0, \quad \text{i.e. } \zeta = 0{\cdot}528.$$

Also $y_m(1) = 0{\cdot}80$, and $y_m(p) \simeq \zeta p$ is a good approximation over the whole range $p > 1$. The values are shown in Fig. 5.15.

Since $\operatorname{re} D(iy, \lambda)$ is an even function of y it is convenient to consider positive values of y. By eqn (5.69)

$$\frac{\mathrm{d}}{\mathrm{d}y} \operatorname{re} D(iy, \lambda) = \int_{p_1}^{p_2} \frac{\partial R(y, z')}{\partial y}\, \sigma(z')\, \mathrm{d}z'.$$

Therefore $$\frac{\mathrm{d}}{\mathrm{d}y}\,\mathrm{re}\,D(iy,\lambda) < 0, \quad \text{for } 0 < y < y_m(p_1), \tag{5.76}$$

$$> 0, \quad \text{for } y_m(p_2) < y < \infty. \tag{5.77}$$

Thus the second condition in eqns (5.75) cannot be fulfilled for $y > y_m(p_2)$. This gives Theorem 12.

THEOREM 12. *The resonance position s_R cannot exceed*

$$s_0 + 4y_m^2(p_2) \simeq s_0 + 4\zeta^2 p_2^2$$
$$\simeq s_0 + 1\cdot 12 p_2^2.$$

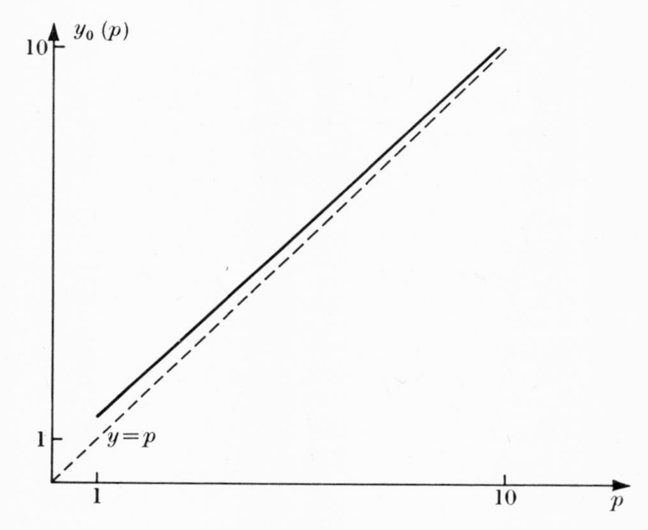

FIG. 5.14. The values of $y_0(p)$.

This result is very different from the e.m.h. case (§ 4.10) in which there exists a λ_R $(0 < \lambda_R < \lambda_1)$ such that s_R decreases from $+\infty$ to s_0 as λ increases from λ_R to λ_1.

Since $$R(y,p) > 0, \quad \text{for } y_0(p) < y \leqslant \infty,$$

it follows that $$\mathrm{re}\,D(iy,\lambda) > 1, \quad \text{for } y_0(p_2) < y \leqslant \infty.$$

Therefore $$0 < \alpha(s) < \tfrac{1}{2}\pi, \quad \text{for } s_0 + 4y_0^2(p_2) < s \leqslant \infty.$$

Also in the range $$s_0 + 4y_m^2(p_2) < s < s_0 + 4y_0^2(p_2)$$

the phase can pass through $\frac{1}{2}\pi$ at most once, and then it must pass downwards through $\frac{1}{2}\pi$. Similarly eqn (5.76) shows that there can be at most one resonance position in

$$s_0 < s < s_0 + 4y_m^2(p_1).$$

However, if the interval $[y_m(p_1), y_m(p_2)]$ is sufficiently large we can construct examples in which there are two resonance positions in

$$s_0 < s < s_0 + 4y_m^2(p_2).$$

For $0 < y < y_0(p_1)$ we have

$$\int_{p_1}^{p_2} R(y, z')\sigma(z')\, dz' < 0,$$

and as λ increases from 0 to λ_1, re $D(iy, \lambda)$ decreases monotonically from 1 to $-\infty$, for any y in this interval. Thus there exists a λ_m $(0 < \lambda_m < \lambda_1)$

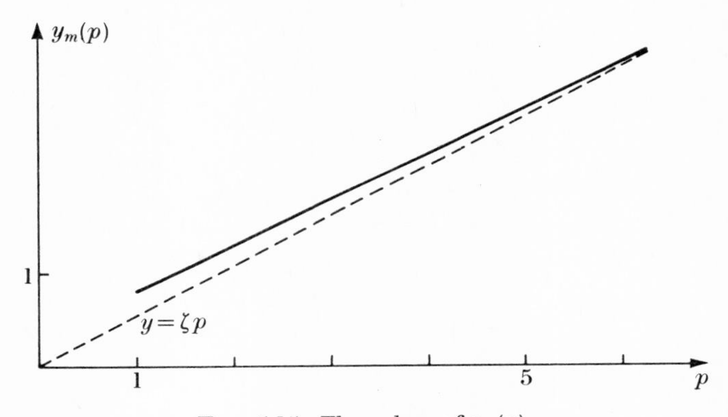

FIG. 5.15. The values of $y_m(p)$.

such that there is a resonance position s_R at $s_0 + 4y_m^2(p_1)$. Equation (5.76) shows that this s_R decreases monotonically to s_0 as λ increases from λ_m to λ_1.

Both eqn (5.56) and Theorem 12 are the optimum results unless we further specify $\bar{\rho}(z)$ for $p_1 \leqslant z \leqslant p_2$. If $\bar{\rho}(z) \propto \delta(z - p_2)$ the maximum value of re $\tilde{s}$ is $s_0 + 4m(p_2)$, and the maximum resonance position is $s_0 + 4y_m^2(p_2) \simeq s_0 + 1{\cdot}12p_2^2$. Using Fig. 5.7 (b) it is clear that if $p \gg 1$ there are values of λ $(0 < \lambda < \lambda_1)$ such that the resonance position does lie far from the pole position $\tilde{s}$.

6

APPROXIMATE CALCULATION OF A P-WAVE PION–NUCLEON PROBLEM

6.1. Introduction

THE general problem of the location of the second sheet singularities of a πN partial wave is a good deal more complicated than the simple equal mass cases we studied in Chapters 4 and 5. The differences arise essentially from the unequal mass kinematics and from the spin of the nucleon.

The unequal mass kinematics imply that in place of rational functions (e.m.h. case, Chapter 4) or circular functions (π–π case, Chapter 5) an exact treatment of the πN problem would involve elliptic functions. The spin of the nucleon means that there are two invariant amplitudes $A(s,t)$ and $B(s,t)$ in πN scattering, and we have to re-examine how to specify a useful problem (a question originally raised in Chapter 1). Let $\Delta f_l(s)$ be the discontinuity in the p.w.a. $f_l(s)$ across the left-hand cut at s. A detailed discussion may be found in [13], p. 186 ff., but it turns out that across those parts of the left-hand cut which are not far away from the physical threshold $s_0 = (M+\mu)^2$, where M and μ are respectively the nucleon and pion masses, $\Delta f_l(s)$ can be expressed in terms of simple t-channel and u-channel exchange processes (e.g. ρ-exchange, σ-exchange, N-exchange, N^*_{33}-exchange).†

From knowledge of $\Delta f_l(s)$ across these portions of the left-hand cut we have what we may call the long and medium range parts of the πN interaction. If we give $\Delta f_l(s)$ across these portions of the left-hand cut (and $\Delta f_l(s) = 0$ on the rest of the left-hand cut) we have specified a useful problem; it is relevant to any πN partial waves in which the short range part of the interaction plays only a minor role. (The role of the short range interaction in various πN partial wave amplitudes is discussed elsewhere [27].)

In order to avoid complications we shall not treat the general problem. Instead we shall give an approximate solution to a πN P-wave whose driving force is an attractive pole at $s = M^2$. Such a model is closely

† Exceptions to this are the cuts $-\infty < s < 0$, and the portion $\arg(s) \gtrsim 66°$ on $|s| = M^2-\mu^2$. For $-(M^2-\mu^2) < s < 0$, however, see J. Lyng Petersen [26].

related to Chew and Low's theory of the N^*_{33} resonance [28]. For simplicity we give only the elastic case, but that restriction could be readily removed.

6.2. The equation for $D(z)$

In this simplified problem the left-hand cut is represented by the pole

$$\frac{C}{s-M^2} \quad (C > 0),$$

in the reduced P-wave amplitude $F(s) = f_1(s)/q^2$. We write

$$F(s) = N(s)/D(s),$$

and by eqn (2.83)

$$N(s) = \frac{CD_p}{s-M^2}, \tag{6.1}$$

where $D_p = D(s = M^2)$. Equation (2.84) gives $D(s)$.

We shall rewrite these equations in terms of the variable

$$z = \tfrac{1}{2}(s_0-s)^{\frac{1}{2}}, \tag{6.2}$$

where z is defined as in Fig. 4.1 (p. 55). Let

$$\left.\begin{aligned} p &= \tfrac{1}{2}(s_0-M^2)^{\frac{1}{2}} \\ p_1 &= \tfrac{1}{2}(s_0-s_1)^{\frac{1}{2}} \\ p_2 &= \tfrac{1}{2}s_0^{\frac{1}{2}} \end{aligned}\right\}, \tag{6.3}$$

where $s_0 = (M+\mu)^2$ and $s_1 = (M-\mu)^2$. Using $M = 6{\cdot}727\mu$ we have the numerical values (in units of μ)

$$p = 1{\cdot}90, \quad p_1 = 2{\cdot}59, \quad \text{and} \quad p_2 = 3{\cdot}86. \tag{6.4}$$

The momentum $q(s)$ is given by eqn (3.1), so

$$q^2 = z^2\frac{z^2-p_1^2}{p_2^2-z^2}. \tag{6.5}$$

The physical cut $s_0 \leqslant s \leqslant \infty$ is the line $\operatorname{re} z = 0$, and in order to have the correct form of $q(s)$ on the physical cut we must use the branch with

$$q(iy) > 0 \quad \text{for } 0 > y > -\infty \tag{6.6}$$

in eqn (6.5). Thus $q(z)$ is regular in the z-plane cut along $-p_2 \leqslant z \leqslant -p_1$ and $p_1 \leqslant z \leqslant p_2$. The main features of $q(z)$ in the cut z-plane are shown in Fig. 6.1. In Fig. 6.2 the values of s corresponding to some points in the z-plane are shown.

By the same methods as we used in Chapters 4 and 5, the unitarity relation gives

$$D(-z) = D(z)+2i(q(z))^3N(z) \tag{6.7}$$

for any z. Also $N(z)$ is an even function of z, and it is regular in the whole z-plane except for simple poles at $z = p$ and $z = -p$. The function $D(z)$

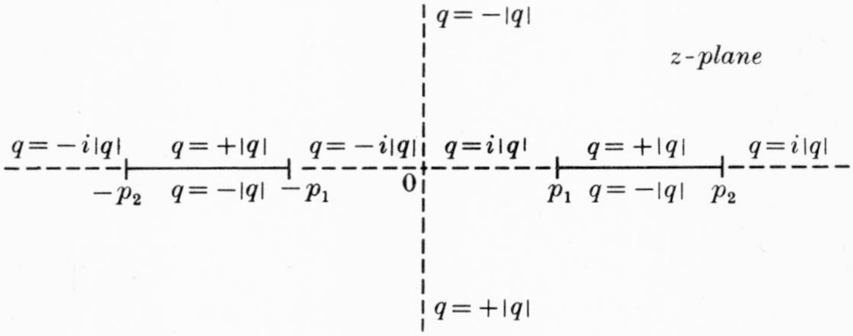

FIG. 6.1. The form of $q(z)$ in the cut z-plane.

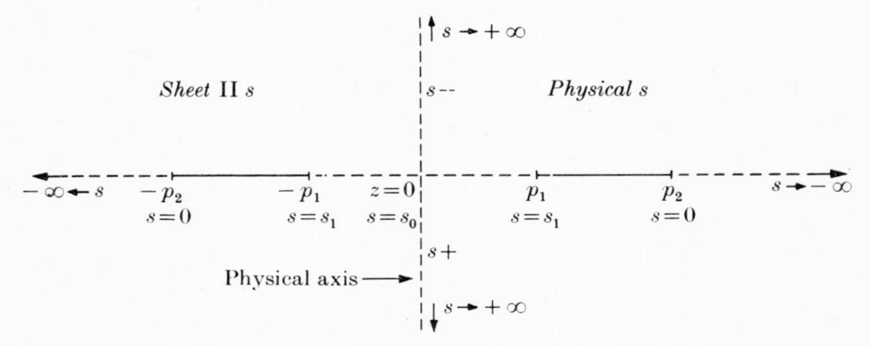

FIG. 6.2. Values of s corresponding to some points in the z-plane.

is regular in the z-plane cut along $-p_2 \leqslant z \leqslant -p_1$, except for a simple pole at $z = -p$.

On $\operatorname{re} z = 0$, $\operatorname{im} D(z) = O(|z|^3)$, so, near $z = 0$,

$$D(z) = 1 + c_2 z^2 + c_3 z^3 + \ldots, \tag{6.8}$$

where $c_2, c_3, \ldots$ are real constants. We can write a dispersion relation for $\big(D(z)-1\big)/z^2$. It is

$$\frac{D(z)-1}{z^2} = \frac{1}{2\pi i} \int\limits_{(\mathscr{C}_1+\mathscr{C}_2)} \mathrm{d}z' \, \frac{D(z')-1}{z'^2(z'-z)}, \tag{6.9}$$

where $\mathscr{C}_1$ and $\mathscr{C}_2$ are the contours shown in Fig. 6.3, and z lies outside these contours. Using eqns (6.6) and (6.7) we can replace $D(z') - 1$ by $-2iq^3(z')N(z')$ in the integral, so

$$\frac{D(z)-1}{z^2} = -\frac{1}{\pi} \int\limits_{(\mathscr{C}_1+\mathscr{C}_2)} \frac{q^3(z')N(z')\,\mathrm{d}z'}{z'^2(z'-z)}. \tag{6.10}$$

The integrals over $\mathscr{C}_1$ and $\mathscr{C}_2$

By eqn (6.1),

$$N(z) = \frac{\tfrac{1}{4}CD_p}{p^2-z^2}, \tag{6.11}$$

hence
$$-\frac{1}{\pi}\int_{\mathscr{C}_2}\frac{q^3(z')N(z')\,\mathrm{d}z'}{z'^2(z'-z)} = \frac{\frac{1}{4}CD_p}{z+p}\,(1-s_1/M^2)^{\frac{3}{2}}. \tag{6.12}$$

We write
$$I_1(z) = -\frac{1}{\pi}\,(\tfrac{1}{4}CD_p)^{-1}\int_{\mathscr{C}_1}\frac{q^3(z')N(z')\,\mathrm{d}z'}{z'^2(z'-z)} = \frac{1}{\pi}\int_{\mathscr{C}_1}\frac{q^3(z')\,\mathrm{d}z'}{z'^2(p^2-z'^2)(z-z')}. \tag{6.13}$$

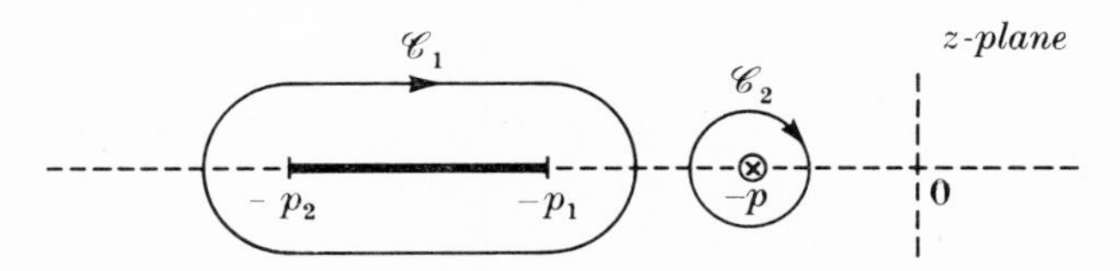

FIG. 6.3. The contours $\mathscr{C}_1$ and $\mathscr{C}_2$ used in eqn (6.9).

A complication in evaluating this integral arises from the fact that the integrand behaves like $(z'+p_2)^{-\frac{3}{2}}$ for small $|z'+p_2|$.

The relation
$$\frac{1}{z-z'} = \sum_{n=0}^{N-1}\frac{(z'+p_2)^n}{(z+p_2)^{n+1}}+\frac{1}{(z+p_2)^N}\frac{(z'+p_2)^N}{z-z'}$$

gives
$$I_1(z) = \sum_{n=0}^{N-1}\frac{a_n}{(z+p_2)^{n+1}}+\frac{R_N(z)}{(z+p_2)^N}, \tag{6.14}$$

where
$$a_n = \frac{1}{\pi}\int_{\mathscr{C}_1}\frac{q^3(z')(z'+p_2)^n}{z'^2(p^2-z'^2)}\,\mathrm{d}z', \quad \text{for } n = 0, 1, 2,\ldots,$$

and
$$R_N(z) = \frac{1}{\pi}\int_{\mathscr{C}_1}\frac{q^3(z')(z'+p_2)^N}{z'^2(p^2-z'^2)(z-z')}\,\mathrm{d}z', \quad \text{for } N = 1, 2,\ldots.$$

For $n \geqslant 1$ the contour $\mathscr{C}_1$ can be shrunk on to the cut $-p_2 \leqslant z' \leqslant -p_1$ without difficulty. This gives
$$a_n = -\frac{2}{\pi}\int_{p_1}^{p_2}\left|\frac{z'^2-p_1^2}{p_2^2-z'^2}\right|^{\frac{3}{2}}\frac{z'(p_2-z')^n}{(z'^2-p^2)}\,\mathrm{d}z', \quad \text{for } n \geqslant 1. \tag{6.15}$$

These are elliptic integrals. It is clear that
$$\left.\begin{aligned} a_n &\leqslant 0 \\ \left|\frac{a_{n+1}}{a_n}\right| &< (p_2-p_1) \end{aligned}\right\} \quad (n \geqslant 1).$$

Substituting $w = z'^2$, we can write

$$a_0 = \frac{1}{2\pi} \int_{\Gamma} \left(\frac{w-p_1^2}{p_2^2-w}\right)^{\frac{3}{2}} \frac{dw}{p^2-w}, \tag{6.16}$$

where the contour Γ is shown in Fig. 6.4. This contour surrounds the cut $p_1^2 \leqslant w \leqslant p_2^2$. The function

$$\left(\frac{w-p_1^2}{p_2^2-w}\right)^{\frac{3}{2}}$$

is defined to be real and positive just above the cut, and it has the value

$$-i\left|\frac{w-p_1^2}{p_2^2-w}\right|^{\frac{3}{2}}$$

for real w outside the cut.

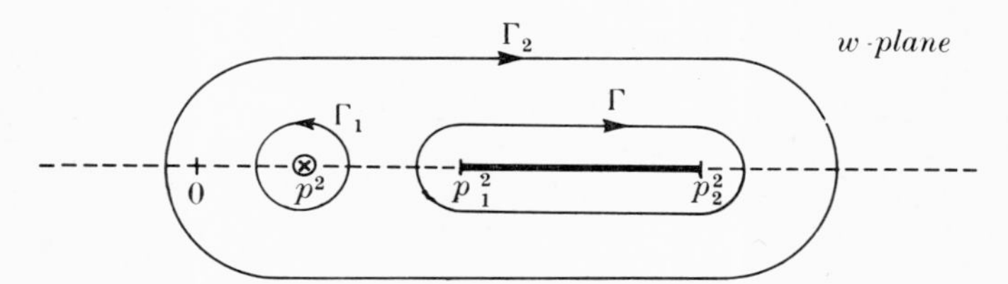

FIG. 6.4. The contours used in eqns (6.16) and (6.17).

Clearly
$$a_0 = \frac{1}{2\pi} \int_{(\Gamma_1+\Gamma_2)} \left(\frac{w-p_1^2}{p_2^2-w}\right)^{\frac{3}{2}} \frac{dw}{p^2-w}, \tag{6.17}$$

where the contours Γ_1 and Γ_2 are shown in Fig. 6.4. Now

$$\frac{1}{2\pi i} \int_{\Gamma_1} \left(\frac{w-p_1^2}{p_2^2-w}\right)^{\frac{3}{2}} \frac{dw}{w-p^2} = -i(1-s_1/M^2)^{\frac{3}{2}}.$$

Also
$$\frac{1}{2\pi i} \int_{\Gamma_2} \left(\frac{w-p_1^2}{p_2^2-w}\right)^{\frac{3}{2}} \frac{dw}{w-p^2} = +i,$$

since
$$\left(\frac{w-p_1^2}{p_2^2-w}\right)^{\frac{3}{2}} (w-p^2)^{-1} \to -i/w, \quad \text{as } |w| \to \infty.$$

Thus eqn (6.17) gives
$$a_0 = 1-(1-s_1/M^2)^{\frac{3}{2}}. \tag{6.18}$$

The values of the first few coefficients a_n (using the parameters in eqn (6.4)) are given in Table 6.1.

TABLE 6.1

Values of the coefficients a_n

a_0	0·856
a_1	—0·350
a_2	—0·085
a_3	—0·044
a_4	—0·030

For $N \geqslant 1$,

$$R_N(z) = -\frac{2}{\pi}\int\limits_{p_1}^{p_2} \left|\frac{z'^2-p_1^2}{p_2^2-z'^2}\right|^{\frac{3}{2}} \frac{(p_2-z')^N z'}{(z'^2-p^2)(z'+z)}\,dz'. \tag{6.19}$$

Hence
$$|R_N(z)| \leqslant \frac{|a_N|}{\min\limits_{z'\in[p_1,p_2]} |z+z'|}. \tag{6.20}$$

6.3. A first approximation to $D(z)$

By eqns (6.10), (6.12), (6.14), and (6.18)

$$D(z) = 1+A\,\frac{z^2}{z+p_2}+B\,\frac{z^2}{z+p}+\tfrac{1}{4}CD_p\,R_1(z)\,\frac{z^2}{z+p_2}, \tag{6.21}$$

where
$$\left.\begin{aligned} A &= \tfrac{1}{4}CD_p\{1-(1-s_1/M^2)^{\frac{3}{2}}\} \\ B &= \tfrac{1}{4}CD_p(1-s_1/M^2)^{\frac{3}{2}} \end{aligned}\right\}, \tag{6.22}$$

and
$$B/A = 0{\cdot}169.$$

Equation (6.19) shows that $R_1(z)$ is real for real z, except for $-p_2 \leqslant z \leqslant -p_1$. Hence the real axis, except for $-p_2 \leqslant z \leqslant -p_1$, is a branch of $\operatorname{im} D(z) = 0$.

The first approximation to $D(z)$ is denoted by $\mathscr{D}(z)$. It is given by

$$\mathscr{D}(z) = 1+\frac{Az^2}{z+p_2}+\frac{Bz^2}{z+p}. \tag{6.23}$$

This will be a good approximation in the large region of the z-plane where $|R_1(z)|/a_0$ is small (see eqn (6.20)). Substituting $z = p$ in eqn (6.23) gives

$$\mathscr{D}_p = \frac{1}{1-0{\cdot}168C}. \tag{6.24}$$

Thus the bound state appears when C has been increased to $C_1 = 5{\cdot}9$.

Equation (6.23) is of the same form as we had in the e.m.h. case in Chapter 4 (§ 4.5). Inserting the form $\bar{\rho}(z) = \alpha\delta(z-p_2)+\beta\delta(z-p)$, with $\alpha, \beta > 0$, in eqn (4.36) would give the form in eqn (6.23). Therefore the branch of $\mathrm{im}\,\mathscr{D}(z) = 0$ off the real axis lies between the circles

$$|z+p_2| = p_2 \quad \text{and} \quad |z+p| = p.$$

Since $A \gg B$, the locus $\mathrm{im}\,\mathscr{D}(z) = 0$ will be close to the former circle. Moreover this locus is independent of the strength (C) of the pole. It is shown in Fig. 6.5.

The effective range equation

$$q^3 \cot \alpha(s) = \mathrm{re}\,\mathscr{D}(s)/\mathscr{N}(s), \tag{6.25}$$

where

$$\mathscr{N}(s) = \frac{C\mathscr{D}_p}{s-M^2}, \tag{6.26}$$

shows that the scattering length is

$$\begin{aligned} a &= \mathscr{N}(s_0) \\ &= \frac{C\mathscr{D}_p}{s_0-M^2}. \end{aligned} \tag{6.27}$$

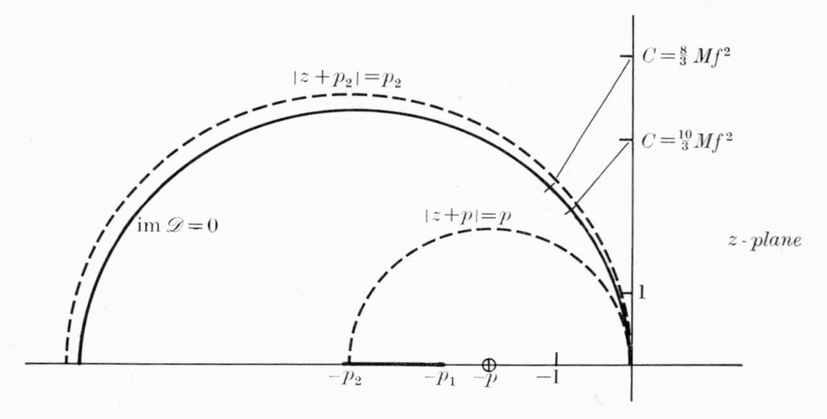

FIG. 6.5. The locus $\mathrm{im}\,\mathscr{D}(z) = 0$ given by eqn (6.23) and the circles $|z+p_2| = p_2$, $|z+p| = p$. The positions $\tilde{z}$ of the pole and the resonance positions iz_R corresponding to $C = \frac{8}{3}Mf^2$ and $\frac{10}{3}Mf^2$ are shown.

On the physical axis $\mathrm{re}\, z = 0$ we write $z = \pm i|z|$. By eqn (6.23),

$$\mathrm{re}\,\mathscr{D}(\pm i|z|) = 1-|z|^2\left\{\frac{Ap_2}{|z|^2+p_2^2}+\frac{Bp}{|z|^2+p^2}\right\}.$$

Now A and B are monotonically increasing functions of C in $0 < C < C_1$. In this range there is a value C_R such that

$$Ap_2+Bp = 1.$$

In fact $C_R = 0{\cdot}94$.

For $0 < C < C_R$ the phase $\alpha(s)$ does not pass through $\frac{1}{2}\pi$ for $s_0 \leqslant s \leqslant \infty$, but $\alpha(s) \to \frac{1}{2}\pi$ as $s \to +\infty$.

For $C_R < C < C_1$, $\mathrm{re}\,\mathscr{D}$ has a zero at $z = \pm iz_R$, so the phase passes through $\frac{1}{2}\pi$ and there is a resonance. Again $\alpha(s) \to \frac{1}{2}\pi$ as $s \to +\infty$ (see § 4.10).

As in § 4.10, if there is a resonance, the position $\tilde{z}$ of the pole will obey

$$0 > \mathrm{re}\,\tilde{z} > -\tfrac{1}{2}p_2 = -1{\cdot}93.$$

6.4. The pole position of $D(z)$

We have to estimate the error in the pole position due to using the approximate expression $\mathscr{D}(z)$ of eqn (6.23) in place of the exact expression for $D(z)$ in eqn (6.21).

It is clear from eqn (6.21) that the branches of $\mathrm{im}\,D(z) = 0$ are independent of C. We shall now show that one branch of $\mathrm{im}\,D(z) = 0$ lies somewhat outside the branch $\mathscr{C}$ of $\mathrm{im}\,\mathscr{D}(z) = 0$ off the real axis. Equation (6.19) shows that the branch of

$$\mathrm{im}\left(\frac{z^2 R_1(z)}{z+p_2}\right) = 0$$

off the real axis must lie between the circles

$$|z+\tfrac{1}{2}p_2| = \tfrac{1}{2}p_2 \tag{6.28}$$

and

$$\left|z+\frac{p_1 p_2}{p_1+p_2}\right| = \frac{p_1 p_2}{p_1+p_2}. \tag{6.29}$$

Also

$$\mathrm{im}\,z\,.\,\mathrm{im}\left(\frac{z^2 R_1(z)}{z+p_2}\right) < 0$$

outside the circle in eqn (6.28). The branch $\mathscr{C}$ of $\mathrm{im}\,\mathscr{D}(z) = 0$ lies outside this circle (see Fig. 6.5), and $\mathrm{im}\,z\,.\,\mathrm{im}\,\mathscr{D}(z) > 0$ for z off the real axis and outside $\mathscr{C}$. This proves the result.

Let z_1 be the complex zero of $\mathscr{D}(z)$ and z_2 the corresponding zero of $D(z)$. Then

$$\Delta z = z_2 - z_1$$

$$\simeq -\frac{z_1^2 R_1(z_1)\frac{1}{4}CD_p}{z_1+p_2} \bigg/ \frac{\mathrm{d}D(z_1)}{\mathrm{d}z}.$$

Now z_1 is near the circle

$$|z+p_2| = p_2.$$

The dominant term in $\mathrm{d}D/\mathrm{d}z$ near z_1 is

$$A\,\frac{\mathrm{d}}{\mathrm{d}z}\left(\frac{z^2}{z+p_2}\right).$$

Therefore

$$\Delta z \simeq -\frac{R_1(z_1)}{a_0}\,\frac{z_1(z_1+p_2)}{z_1+2p_2}. \tag{6.30}$$

Let $\theta = \arg(z_1+p_2)$. Equation (6.20) shows that for $0 < \theta < \bar{\theta}$ where

$$\bar{\theta} = \arccos\{1-p_1/p_2\} \simeq 70°,$$

eqn (6.30) gives

$$|\Delta z| \leqslant \frac{|a_1|}{a_0} \frac{p_2 \tan(\frac{1}{2}\theta)}{\{p_1^2+4p_2(p_2-p_1)\sin^2(\frac{1}{2}\theta)\}^{\frac{1}{2}}}. \tag{6.31}$$

For $\theta > \bar{\theta}$,

$$|\Delta z| \lesssim \frac{|a_1|}{a_0} \tan(\tfrac{1}{2}\theta). \tag{6.32}$$

For $0 < \theta \lesssim \frac{1}{4}\pi$, eqn (6.31) gives

$$|\Delta z| \lesssim 0{\cdot}3\theta, \tag{6.33}$$

and even for $\theta = \frac{1}{3}\pi$, $|\Delta z|$ is no greater than about 0·25.

Comparing with Fig. 6.5, it is seen that z_1 is a good approximation to z_2 provided θ is not large (i.e. $\theta \not> \frac{1}{2}\pi$). Thus the approximation $\mathscr{D}(z)$ (eqn (6.23)) gives a good estimate of the position of the second sheet pole if $\theta \lesssim \frac{1}{2}\pi$. Similarly it gives a good estimate of the resonance position iz_R.

6.5. Behaviour of $D(z)$ near the cut

Because of the last term on the right of eqn (6.21), $\operatorname{im} D(z) = 0$ may have another branch off the real axis. This could only lie inside the circle in eqn (6.28).

By eqn (6.7)

$$\operatorname{im} D(z) = \operatorname{im} D(-z) - \operatorname{im}\big(2iq^3(z)N(z)\big). \tag{6.34}$$

From eqn (6.11) and Fig. 6.1, it follows that

$$\operatorname{im} D(z+) = \frac{|q(z)|^3 \frac{1}{2}CD_p}{z^2-p^2}, \quad \text{for } -p_2 \leqslant z \leqslant -p_1,$$

so $\operatorname{im} D > 0$ just above the cut $-p_2 \leqslant z \leqslant -p_1$.

It can be shown that

$$\operatorname{re}\big(2iq^3(z)N(z)\big)$$

is an increasing function of z on $-p_1 \leqslant z \leqslant 0$, therefore

$$\operatorname{im}\big(2iq^3(z)N(z)\big) > 0$$

just above the interval $-p_1 < z < 0$. Since $\operatorname{im} D(-z) < 0$ when z is just above the interval $-p_1 < z < 0$, it follows from eqn (6.34) that the same is true for $\operatorname{im} D(z)$.

These results show that branches of $\operatorname{im} D(z) = 0$ cannot leave the interval $-p_2 \leqslant z < 0$ of the real axis except at the points $-p_2$ and $-p_1$. It follows from eqns (6.5) and (6.34) that a branch of $\operatorname{im} D = 0$ leaves the real axis at $z = -p_1$, and it is tangential to the real axis there. Since $D(z)$ is proportional to $(z+p_2)^{-\frac{3}{2}}$ near $z = -p_2$, three branches of $\operatorname{im} D = 0$ meet at $z = -p_2$ with angular separation $\frac{2}{3}\pi$ (see § 3.5).

Thus there is a branch of $\operatorname{im} D = 0$ inside the circle of eqn (6.28) and it must have the form shown in Fig. 6.6.

Finally we can show that, for $0 < C < C_1$, there is no zero of $D(z)$ lying off the real axis and on this branch of $\operatorname{im} D = 0$. Clearly $\operatorname{im} D < 0$ just above the portion $(-p_2)Q(-p_1)(-p)$ of the branch of $\operatorname{im} D = 0$ (see Fig. 6.6). Therefore $\operatorname{re} D$ decreases as we move from $-p_2$ to $-p$ along this portion of the branch. In fact $\operatorname{re} D$ decreases from $+\infty$ to $-\infty$, so there is a zero of $D(z)$ on the branch $(-p_2)Q(-p_1)(-p)$.

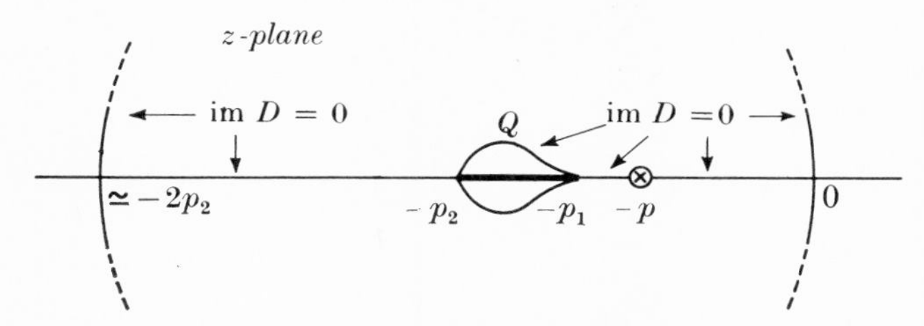

FIG. 6.6. The branches of im $D = 0$ near and on the real axis.

However, on evaluating $R_1(-p_1)$ (eqn (6.19)) we find from eqn (6.21) that

$$D(-p_1) = 1+2{\cdot}62CD_p.$$

Hence for $0 < C < C_1$, $D(z)$ has a zero on $-p_1 \leqslant z \leqslant -p$, but it has no zero on the branch $(-p_2)Q(-p_1)$ of $\operatorname{im} D = 0$.

It follows from the results in §§ 6.3, 6.4, and 6.5 that for $0 < C < C_1$, $D(z)$ has at most one complex conjugate pair of zeros off the real axis.

6.6. Practical examples

Consider the πN amplitude P_{33}. The dominant driving forces are known (p. 284 ff. of [29]) to be the long range N-exchange interaction and the σ-exchange interaction; the former is much the stronger.

This long range N-exchange is well represented by the pole

$$\frac{\frac{8}{3}Mf^2}{s-M^2} \tag{6.35}$$

in the reduced p.w.a. $F(s)$. Here $f^2 = 0{\cdot}081$ is the πN coupling constant. Suppose that this is the only interaction. Then the results of § 6.3 hold with $C = \frac{8}{3}Mf^2 = 1{\cdot}45$. Then by eqns (6.22),

$$A = 0{\cdot}41, \qquad B = 0{\cdot}069.$$

The pole position is

$$\tilde{z} = -1{\cdot}11 + i\,2{\cdot}50$$

(see Fig. 6.5), and the resonance position is $z_R = 4{\cdot}35$, corresponding to $s_R = 135{\cdot}5$. Notice that iz_R and $\tilde{z}$ are far apart.

This solution has $\mathscr{D}_p = 1{\cdot}32$ and the scattering length is $a = 0{\cdot}134$. Thus the interaction in eqn (6.35) gives the resonance position at a high energy compared with the observed value $(s_R)_{\text{exp}} \simeq 80$, and it gives the scattering length much smaller than the experimental value $(a)_{\text{exp}} \simeq 0{\cdot}215$.

This disagreement is partly due to ignoring the σ-exchange interaction, partly to ignoring various smaller interactions of shorter range, and partly due to ignoring the inelasticity which appears in the P_{33} amplitude at high energies. For further discussion of this topic see p. 245 of [29].

An improved model

Here we shall only examine the improvement arising from a rough method of including σ-exchange, described in [29], p. 242, namely we increase C by 25%, i.e. we use $C = \frac{10}{3}Mf^2 = 1{\cdot}82$ in eqn (6.35). This gives

$$A = 0{\cdot}56, \qquad B = 0{\cdot}094.$$

The pole position becomes

$$\tilde{z} = -0{\cdot}82 + i\,2{\cdot}18$$

(see Fig. 6.5) and the resonance position is $z_R = 3{\cdot}16$, corresponding to $s_R = 99{\cdot}7$. This solution gives $\mathscr{D}_p = 1{\cdot}44$ and $a = 0{\cdot}18$.

The resonance position and the scattering length are much closer to the experimental values than in the preceding example, but because of the other features which have been ignored we do not expect to get good agreement with the experimental values (see [29], p. 245). The errors arising from using $\mathscr{D}(z)$ instead of $D(z)$ are small.

Remark

In conclusion we emphasize that the effect of the πN kinematics has been to move the pole position $\tilde{z}$ near to the circle $|z+p_2| = p_2$, instead of being on the circle $|z+p| = p$, where it should be in an e.m.h. case.

7

PATHOLOGICAL AMPLITUDES

7.1. Introduction

THERE are two troublesome questions: (*a*) how is a resonance to be defined? (*b*) what is the relation between unstable particles and S-matrix poles? Some discussion of question (*b*) has been given by Calucci *et al.* [30], [31], with particular emphasis on constructing partial wave amplitudes which have no second sheet singularities away from the real axis, and on the relation of such amplitudes to the exponential decay law [31].

In the preceding chapters we have defined a resonance by the property that (in the elastic case) the phase $\alpha(s)$ passes up through $\frac{1}{2}\pi$, and the energy where this occurs is taken as the resonance energy $s_R^{\frac{1}{2}}$. However unsatisfactory this definition may be from some other viewpoints, it is certainly useful when studying the properties of partial wave dispersion relations.

It is possible that we shall run into trouble if we try to define a resonance by the second sheet poles of the p.w.a. We saw in § 4.10 in the case of e.m.h. P-waves that the resonance pole can lie very far away from the resonance position s_R, in the case of broad resonances. Instead of discussing questions (*a*) or (*b*), it is more in the spirit of the present work to examine the related phenomenon of *pathological partial wave amplitudes*. These are defined as (elastic) amplitudes having a resonant phase shift and having no second sheet singularities away from the real axis $-\infty \leqslant s < s_0$ (s_0 is always the physical threshold). Among the examples given by Calucci *et al.* [30] are some having narrow resonances, and also some examples having essential singularities on the physical sheet. Various other types are known, and it seems profitable to survey them briefly.

7.2. The two pole e.m.h. P-wave

In Chapter 4 we examined the e.m.h. P-wave problem having a left-hand cut $p_1 \leqslant z \leqslant p_2$ (with $z = (s_0-s)^{\frac{1}{2}}$). As the strength parameter λ is increased from zero, the second sheet poles move along the real axis $-\infty < s < s_2$, and at a critical value λ_c ($0 < \lambda_c < \lambda_1$) two poles meet and leave the real axis (§ 4.9). If $p_2 > 2p_1$, these poles could return to the real axis for some range(s) of values of λ in $\lambda_c < \lambda < \lambda_1$ and z in $-p_2 \leqslant z \leqslant -2p_1$. In § 4.10 we examined the further restriction imposed by requiring that there be a resonance (i.e. that $\alpha(s)$ passes up through $\frac{1}{2}\pi$). Then the second sheet pole position $\tilde{z}$ will obey $\operatorname{re}\tilde{z} \geqslant -\frac{1}{2}p_2$. Thus if $p_2 < 4p_1$, $\tilde{z}$ cannot lie on the real axis, but we cannot exclude the possibility that this occurs for $p_2 > 4p_1$.

In this section we shall show that in the simple case where the driving force consists of two poles, and there is a resonance, it is possible for all the second sheet poles to lie on the real axis.

In the notation used in eqn (4.36) we put

$$\bar{\rho}(z) = \Gamma_1\,\delta(z-p_1)+\Gamma_2\,\delta(z-p_2),$$

where Γ_1 and Γ_2 are positive. By eqn (4.36),

$$D(z,\lambda) = 1+\frac{z^2\gamma_1}{z+p_1}+\frac{z^2\gamma_2}{z+p_2}, \tag{7.1}$$

where
$$\gamma_i = \frac{\lambda}{4\pi}\Gamma_i p_i D(p_i,\lambda) \quad (i = 1,2).$$

Since $0 < \lambda < \lambda_1$, it follows that γ_1 and γ_2 are positive. Putting $z = p_i$ in eqn (7.1) gives

$$\gamma_i = \Gamma'_i+\Gamma'_i p_i^2 \sum_{j=1}^{2} \frac{\gamma_j}{p_i+p_j} \quad (i = 1,2), \tag{7.2}$$

with
$$\Gamma'_i = \left(\frac{\lambda}{4\pi}\right)p_i\Gamma_i \quad (i = 1,2).$$

When the determinant of eqns (7.2) vanishes, i.e.

$$\left|\delta_{ij}-\frac{\Gamma'_i p_i^2}{p_i+p_j}\right| = 0 \quad (i,j = 1,2),$$

there is a bound state at the physical threshold s_0. Thus the region $\mathcal{N}$ in the (Γ'_1,Γ'_2)-plane in which there are no bound states is bounded by the axes $\Gamma'_1 = 0$, $\Gamma'_2 = 0$, and a branch of the hyperbola

$$(1-\tfrac{1}{2}p_1\Gamma'_1)(1-\tfrac{1}{2}p_2\Gamma'_2)-\left(\frac{p_1p_2}{p_1+p_2}\right)^2\Gamma'_1\Gamma'_2 = 0 \tag{7.3}$$

(see Fig. 7.1). Moreover the transformation in eqns (7.2) maps this region $\mathcal{N}$ onto the first quadrant of the (γ_1,γ_2)-plane.

From (7.1) the zeros of $D(z,\lambda)$ are given by the roots of the equation

$$\frac{\gamma_1}{z+p_1}+\frac{\gamma_2}{z+p_2}+\frac{1}{z^2} = 0. \tag{7.4}$$

This is a cubic equation in z. It always has a real root in $(-p_2, -p_1)$. If the other two roots are real they will lie in $(-\infty, -p_2)$ if $p_2 < 2p_1$ or in $(-\infty, -2p_1)$ otherwise.

We write $$P = p_1+p_2, \quad Q = p_1 p_2,$$
and in place of γ_1, γ_2 we introduce the real variables (x, y) by
$$yP^3/Q^2 = \gamma_1+\gamma_2,$$
$$xP^2/Q = \gamma_1 p_2+\gamma_2 p_1+1.$$
Then $y > 0$ and $x > Q/P^2$. In this notation eqn (7.4) becomes
$$yP^3z^3+xP^2Qz^2+PQ^2z+Q^3 = 0. \tag{7.5}$$

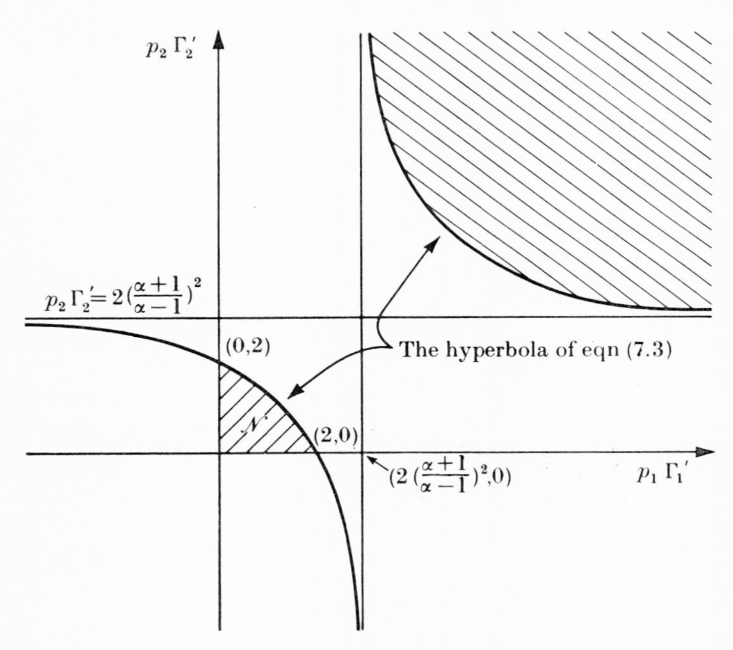

FIG. 7.1. The figure is drawn for $\alpha = p_2/p_1 = 9$. In the region $\mathscr{N}$ there is no bound state, in the unshaded region in the first quadrant there will be one, and in the outer shaded region there will be two bound states.

The discriminant of this cubic is
$$\mathscr{D} = (P^6/Q^2)(-4x^3+x^2+18xy-27y^2-4y). \tag{7.6}$$
If $\mathscr{D} < 0$, eqn (7.5) has one real and two complex roots, and if $\mathscr{D} \geqslant 0$ it has three real roots. Equation (7.6) can be written
$$\mathscr{D} = -27(P^6/Q^2)\{(y-\tfrac{1}{3}x+\tfrac{2}{27})^2-(\tfrac{2}{27})^2(1-3x)^3\}. \tag{7.7}$$
So $\mathscr{D} \geqslant 0$ can only occur for $x \leqslant \frac{1}{3}$. Also the region where $\mathscr{D} > 0$ lies between the curves
$$\mathscr{D}_1 : y = \tfrac{1}{3}x-\tfrac{2}{27}\{1-(1-3x)^{\frac{3}{2}}\} \tag{7.8}$$
and
$$\mathscr{D}_2 : y = \tfrac{1}{3}x-\tfrac{2}{27}\{1+(1-3x)^{\frac{3}{2}}\} \tag{7.9}$$

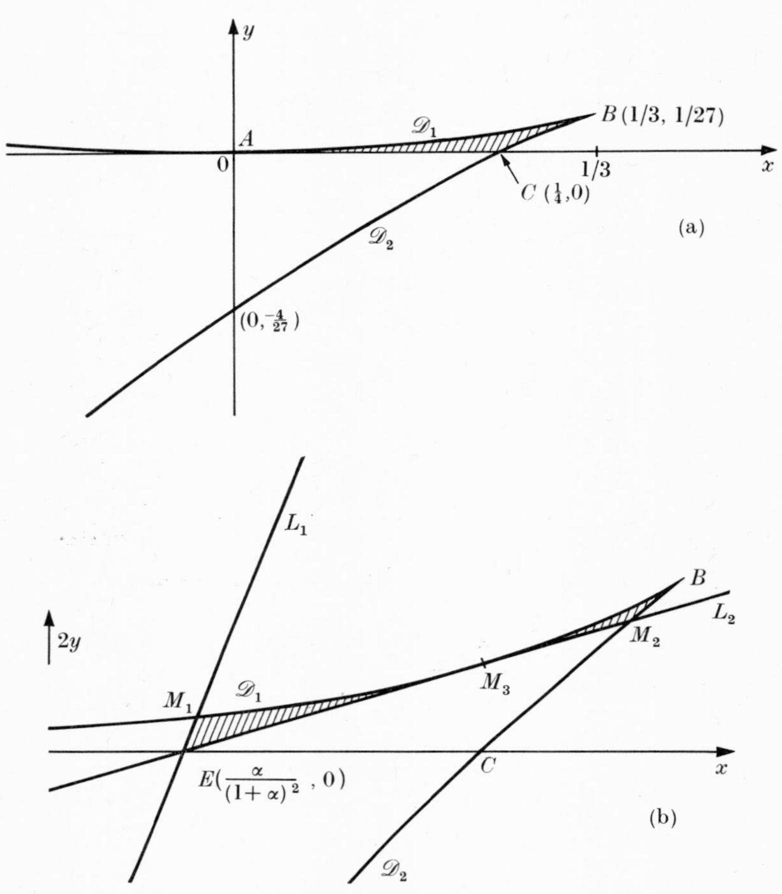

FIG. 7.2 (a). The curves $\mathscr{D}_1$ and $\mathscr{D}_2$. The region in the first quadrant where $\mathscr{D} \geqslant 0$ is shaded. (b) The lines L_1 and L_2 for $\alpha > 2$. The shaded regions correspond to $\mathscr{D} \geqslant 0$ in the first quadrant of (γ_1, γ_2).

(see Fig. 7.2(a)). These curves meet at $B = (\frac{1}{3}, \frac{1}{27})$. $\mathscr{D}_2$ crosses the x-axis at $C = (\frac{1}{4}, 0)$, and $\mathscr{D}_1$ touches the x-axis at $A = (0, 0)$.

Only part of the region ABC corresponds to values in the first quadrant of the (γ_1, γ_2)-plane. The lines $\gamma_1 = 0$ and $\gamma_2 = 0$ are

$$L_1\colon y = \frac{\alpha}{1+\alpha}x - \frac{\alpha^2}{(1+\alpha)^3}, \tag{7.10}$$

$$L_2\colon y = \frac{x}{1+\alpha} - \frac{\alpha}{(1+\alpha)^3}, \tag{7.11}$$

respectively. We have written $p_2 = \alpha p_1$, so $\alpha \geqslant 1$. These lines meet at the point $E = (\alpha/(1+\alpha)^2, 0)$ on the x-axis. As α increases from 1 to ∞,

E moves from C to A. The line L_1 always touches the curve $\mathscr{D}_2$ in $y \leqslant 0$. For $1 \leqslant \alpha \leqslant 2$ the line L_2 touches $\mathscr{D}_2$ on the arc CB, and for $2 \leqslant \alpha \leqslant \infty$ L_2 touches $\mathscr{D}_1$ on the arc BA; these results are easily proved.

The situation for $\alpha > 2$ is shown in Fig. 7.2 (b). There are two regions, $M_1 M_3 E$ and $M_3 BM_2$, which correspond to $\mathscr{D} > 0$ for values in the first quadrant of the (γ_1, γ_2)-plane. As $\alpha \to \infty$ the region $M_3 BM_2$ increases to fill the whole region ABC. When (x, y) is in the region $M_3 BM_2$ the roots of eqn (7.5) all lie in $-p_2 \leqslant z \leqslant -p_1$.

Consider the line

$$\gamma_1 p_1 + \gamma_2 p_2 = \delta, \tag{7.12}$$

where δ is a constant and $\delta \geqslant 1$. It follows from re D (eqn (7.1)) that if $\delta = 1$ there is a resonance at infinity. If $\delta > 1$ the resonance occurs at a finite energy. Equation (7.12) can be written as

$$y = \frac{\alpha}{(1+\alpha)^2}x + (\delta - 1)\frac{\alpha^2}{(1+\alpha)^4}. \tag{7.13}$$

First consider the case $\delta = 1$. Let $(1+\alpha_B)^2 = 9\alpha_B$, and $\alpha_B > 1$, i.e.

$$\alpha_B = \tfrac{1}{2}(7+3\surd 5) \simeq 6{\cdot}85.$$

If $\alpha = \alpha_B$ this line passes through B, and if $\alpha > \alpha_B$ it passes through the shaded region in Fig. 7.2 (b). Therefore for any $\alpha \geqslant \alpha_B$ there exist (positive) values (Γ_1, Γ_2) such that there is a resonance at infinity and the three second sheet poles lie on the real axis $(-\infty < s < s_1)$.

It is clear that for $\alpha > \alpha_B$ there is a range of values of δ $(\delta > 1)$ for which the line in eqn (7.13) passes through the shaded region in Fig. 7.2 (b). Hence for $\alpha > \alpha_B$ there exist values (Γ_1, Γ_2) such that there is a resonance at finite energy, and the second sheet poles all lie on the real axis.

It is fairly easy to show that such resonances can be at moderate energies. Putting $z = iz_R = i(p_1 p_2)^{\frac{1}{2}}$ in eqn (7.1) we find that re $D = 0$ for

$$\gamma_1 + \gamma_2 = P/Q = \frac{1}{p_1}\left(1+\frac{1}{\alpha}\right),$$

i.e. for

$$y = \alpha/(1+\alpha)^2.$$

This line passes through B for

$$(1+\alpha)^2 = 27\alpha,$$

or

$$\alpha \simeq 25.$$

So the resonance is at $z_R \simeq 5p_1$ (i.e. $s_R \simeq s_0 + 25p_1^2$).

7.3. Unfamiliar singularities

Calucci *et al.* [30], [31] have set up classes of pathological amplitudes in which the position where the phase shift goes through $\frac{1}{2}\pi$ is arbitrary, and the width of the resonance can be as small as we like. We shall examine some of their analytic properties.

It is simplest to take the e.m.h. kinematics. For simplicity we write E for the complex variable $\frac{1}{4}(s-s_0)$, so that

$$q^2 = E. \tag{7.14}$$

The physical sheet value

$$q(E) = E^{\frac{1}{2}} \tag{7.15}$$

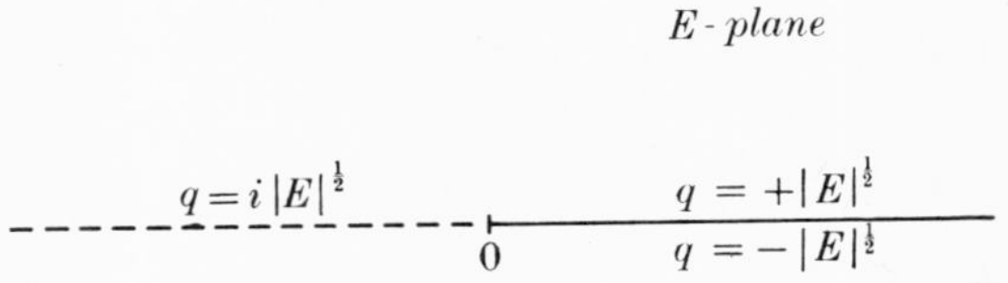

E-plane

$q = i|E|^{\frac{1}{2}}$ $q = +|E|^{\frac{1}{2}}$

0 $q = -|E|^{\frac{1}{2}}$

FIG. 7.3. Definition of q in the (physical sheet) E-plane cut along $0 \leqslant E \leqslant \infty$.

is defined in the E-plane cut along $0 \leqslant E \leqslant \infty$, as in Fig. 7.3. The second sheet values, obtained by passing through the cut, are

$$q^{\text{II}}(E) = -q(E). \tag{7.16}$$

We define the auxiliary functions

$$h_{l,n}(w) = (K_{l,n})^{-1} \int\limits_{-\eta}^{w} g_{l.n}(t)\, \mathrm{d}t, \tag{7.17}$$

where w is complex and $0 < \eta < 1$; also $l = 0, 1, 2,...$ is the orbital angular momentum and n is any positive integer. Further,

$$K_{l,n} = \int\limits_{-\eta}^{0} g_{l,n}(t)\, \mathrm{d}t, \tag{7.18}$$

and

$$g_{l,n}(t) = (1-t^2)^n(1-t^2/\eta^2)^l. \tag{7.19}$$

Since $g_{l,n}(t)$ is a regular function of t, the path of integration in eqn (7.17) does not matter. It is clear that

$$h_{l,n}(-\eta) = 0, \quad h_{l,n}(0) = 1, \quad h_{l,n}(\eta) = 2, \tag{7.20}$$

and

$$h_{l,n}(x) = O((x+\eta)^{l+1}), \quad \text{for } x \simeq -\eta. \tag{7.21}$$

Also $K_{l,n}$ are positive numbers and

$$K_{l,n} = O(n^{-\frac{1}{2}}), \quad \text{as } n \to \infty. \tag{7.22}$$

We have

$$h'_{l,n}(x) > 0, \quad \text{for } -\eta \leqslant x \leqslant \eta,$$

and

$$h'_{l,n}(0) = (K_{l,n})^{-1}. \tag{7.23}$$

For large $|w|$

$$h_{l,n}(w) \simeq (-1)^{n+l}\frac{w^{(2n+2l+1)}}{(2n+2l+1)}(K_{l,n})^{-1}\eta^{-2l}. \tag{7.24}$$

First example

Calucci *et al.* [30] assume that the phase $\delta_l(E)$ for orbital angular momentum l has the form

$$\delta_l(E) = \tfrac{1}{2}\pi\left(\frac{a}{E}\right)^{\frac{1}{2}} h_{l,n}(\phi(E)), \tag{7.25}$$

where $a > 0$, and

$$\phi(E) = \eta\frac{E-a}{E+a}. \tag{7.26}$$

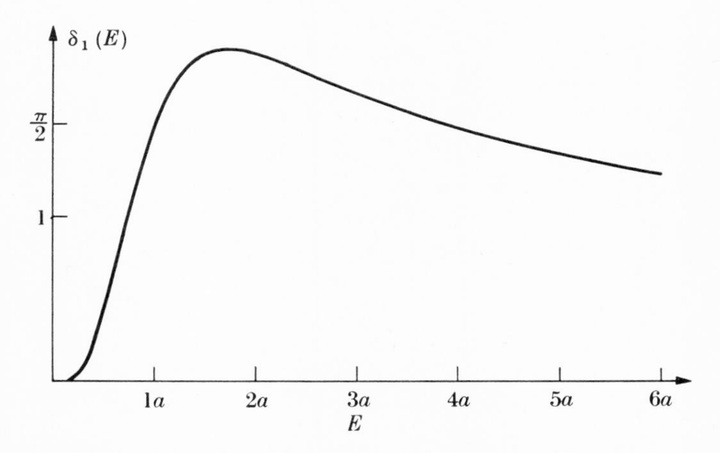

Fig. 7.4. The P-wave phase $\delta_1(E)$ given by eqns (7.25) and (7.26) with $n = 25$ and $\eta = \frac{1}{2}$.

Thus $\delta_l(E)$ is an odd function of $q(E)$, and eqn (7.21) shows that it obeys the correct centrifugal condition $\delta_l(E) \sim q^{2l+1}$ near $E = 0$. By eqn (7.20) $\delta_l = \frac{1}{2}\pi$ at $E = a$, and $\delta_l(E) = O(E^{-\frac{1}{2}})$ as $E \to +\infty$. The phase $\delta_l(E)$ is positive and never reaches π in $0 \leqslant E \leqslant \infty$. For large n the width of the resonance becomes very small, since

$$\left.\frac{\mathrm{d}\delta_l(E)}{\mathrm{d}E}\right|_{E=a} = O(n^{\frac{1}{2}}).$$

An example is shown in Fig. 7.4.

The S-matrix $S_l(E)$ is obtained by substituting eqn (7.25) in

$$S_l(E) = \exp(2i\delta_l(E)). \tag{7.27}$$

Apart from the physical cut $0 \leqslant E \leqslant \infty$, the only singularity of $S_l(E)$

on the physical or the second sheet is at $E = -a$. Using eqns (7.24) and (7.26) we find, on the physical sheet,

$$S_l(E) \simeq \exp\left\{\frac{(-1)^{n+l+1}\pi\eta^{-2l}}{(2n+2l+1)K_{l,n}}\left(\frac{2a\eta}{E+a}\right)^{(2n+2l+1)}\right\}, \quad \text{for } E \simeq -a. \tag{7.28}$$

Also $S_l^{\mathrm{II}}(E) = (S_l(E))^{-1}$. Therefore at $E = -a$ on both sheets the p.w.a. has an essential singularity for all $l \geqslant 0$ and $n \geqslant 0$.

In strong interaction dynamics we do not consider amplitudes having essential singularities (at least on the physical sheet), so the examples given by eqns (7.25) and (7.26) are not to be regarded as physical.

Second example

Calucci *et al.* [30] propose to avoid the difficulty just mentioned by replacing eqns (7.25) and (7.26) by

$$\delta_l(E) = \tfrac{1}{2}\pi\left(\frac{a}{E}\right)^{\frac{1}{2}}\left(\frac{b+E}{b+a}\right)^{\frac{1}{3}} h_{l,n}(\phi(E)) \tag{7.29}$$

and

$$\phi(E) = \frac{\eta}{\ln(1+a/b)}\ln\left(\frac{b+E}{b+a}\right), \tag{7.30}$$

where a and b are positive constants. Also,

$$\left(\frac{b+E}{b+a}\right)^{\frac{1}{3}} \quad \text{and} \quad \ln\left(\frac{b+E}{b+a}\right)$$

are defined in the plane cut along $-\infty \leqslant E \leqslant -b$ (we shall call these the principal values).

The behaviour of $\delta_l(E)$ on $0 \leqslant E \leqslant \infty$ is of the same nature as in the preceding example. Moreover there is no singularity at $E = -a$, but $\delta_l(E)$ has a cut $-\infty \leqslant E \leqslant -b$.

Suppose that $\delta_l(E)$ had a logarithmic branch point at $E = -b$. Since $\delta_l(E)$ has to be an odd function of $E^{\frac{1}{2}}$, and $E^{\frac{1}{2}} = i|E|^{\frac{1}{2}}$ on $-\infty < E < 0$, we must have

$$\delta_l(E) = i\lambda\ln(E+b), \quad \text{near } E = -b,$$

where λ is real. On going N times around the branch point,

$$\delta_l(E) \to \delta_l(E) - 2\pi\lambda N. \tag{7.31}$$

This change does not alter $\operatorname{im}\delta_l(E)$ so $|S_l(E)|$ is not altered.

However, the singularity at $E = -b$ (in eqns (7.29) and (7.30)) is not of the logarithmic type. On the physical sheet or the second sheet we have for $E \simeq -b$,

$$|\delta_l(E)| < C_{n,l}|b+E|^{\frac{1}{3}}\big|\ln|b+E|\big|^{2n+2l+1},$$

where $C_{n,l}$ is a constant depending only on n and l. Thus as $E \to -b$, $\delta_l(E) \to 0$ on either of these sheets.

We can go around the branch point $z = -b$ in the physical or the second sheets. If we go around N times, by eqn (7.30)

$$\phi(E) \to \phi(E) + \frac{2\pi N \eta i}{\ln(1+a/b)}. \tag{7.32}$$

If we choose the original value E so that $|\phi(E)| \gg 1$, then the change in eqn (7.32) induces the change

$$\delta_l(E) \gtrsim \delta_l(E) + (-1)^{n+l}\exp(i\theta)\frac{\pi^2 N}{K_{n,l}}\left(\frac{a}{b}\right)^{\frac{1}{2}}\left(\frac{b+E}{b+a}\right)^{\frac{1}{3}}\frac{(\ln(b+E))^{2(n+l)}}{(\ln(1+a/b))^{2(n+l)+1}}\,\eta^{(2n+1)} \tag{7.33}$$

where $\theta = 0$, $\frac{2}{3}\pi$, or $\frac{4}{3}\pi$, and the principal values of $\ln(b+E)$ and $(b+E)^{\frac{1}{3}}$ are used.

Given z such that $|z+b|$ is as small as we like (but not zero) and given any large number B', there exists an integer N' such that

$$N'|b+z|^{\frac{1}{3}}(\ln|b+z|)^{2(n+l)} > B'.$$

It follows that if $|E+b|$ is a given small number, M is any large positive integer, and B any large positive number, there exists an integer N and a value of E such that

$$\mathrm{im}\left\{(-1)^{n+l}\frac{\pi^2 N}{K_{n,l}}\left(\frac{a}{b}\right)^{\frac{1}{2}}\left(\frac{b+E}{b+a}\right)^{\frac{1}{3}}\frac{(\ln(b+E))^{2(n+l)}}{(\ln(1+a/b))^{2n+2l+1}}\,\eta^{2n+1}\right\} < -\frac{B}{|E+b|^M}.$$

Therefore there is a sequence of points E_j such that $|E_j+b| \to 0$ as $j \to +\infty$ and

$$\ln|S_l(E_j)| > \frac{B}{|E_j+b|^M} \tag{7.34}$$

for all j. This is very different behaviour from what is possible near a logarithmic branch point.

The classification of the singularity structure of multivalued analytic functions is not commonly known, and we shall not discuss it (a useful discussion may be found in § 10.6 of [32]). However, it is clear that eqn (7.34) indicates a type of behaviour which we would like to exclude from the p.w.a. in strong interaction dynamics. On these grounds we conjecture that this second example of Calucci *et al.* [30] is also unphysical.

7.4. The arctan functions†

Simple models of the S-function can be derived from certain rational functions of the momentum. We use the same kinematics as in § 7.3 so

† For the ideas in this section we are indebted to J. Lyng Petersen.

that q is given by eqns (7.14), (7.15), and (7.16). The physical sheet gives $\operatorname{im} q \geqslant 0$ in the q-plane and the second sheet gives $\operatorname{im} q \leqslant 0$. Also

$$q(E^*) = -q^*(E). \tag{7.35}$$

Consider the function

$$S(E) = \left(-\frac{q-ib}{q+ib}\right)^P, \tag{7.36}$$

where P is a real number and b is a positive number. If P is a positive integer the only singularity of $S(E)$ is a pole of order P at $E = -b^2$ on the second sheet, while if P is a negative integer the only singularity is a similar pole at $E = -b^2$ on the physical sheet.

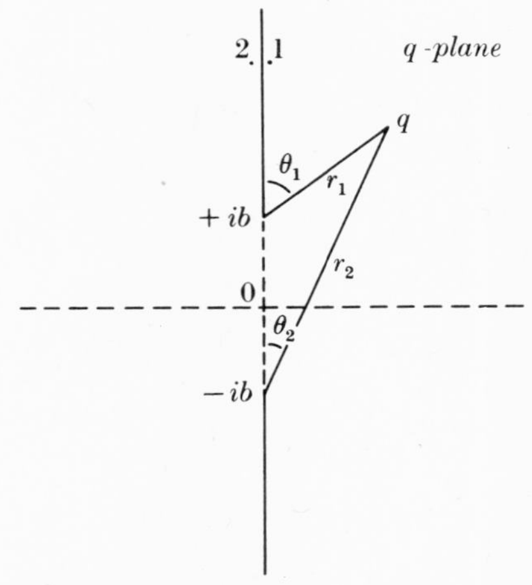

FIG. 7.5. Definition of the function
$$\left(-\frac{q-ib}{q+ib}\right)^P = \left(\frac{r_1}{r_2}\right)^P \exp\{iP(\pi-\theta_1+\theta_2)\} \quad (P \neq \text{integer}).$$

If P is not an integer we cut the q-plane along $(-i\infty, -ib)$ and $(ib, i\infty)$ as in Fig. 7.5. At point 1 we have

$$\left(-\frac{q-ib}{q+ib}\right)^P = \exp(i\pi P)\left(\frac{r_1}{r_2}\right)^P,$$

and at point 2 we have the complex conjugate of this. Thus on the physical sheet for $-\infty < E < -b^2$,

$$S(E\pm) = \exp(\pm i\pi P)\left(\frac{\sqrt{(-E)}-b}{\sqrt{(-E)}+b}\right)^P. \tag{7.37}$$

Similarly on the second sheet for $-\infty < E < -b^2$,

$$S^{\mathrm{II}}(E\pm) = \exp(\mp i\pi P)\left(\frac{\sqrt{(-E)}+b}{\sqrt{(-E)}-b}\right)^P. \tag{7.38}$$

These equations give the only singularities when P is not an integer.

The upper side of the physical cut $0 \leqslant E \leqslant \infty$ corresponds to $0 \leqslant q \leqslant \infty$. Equation (7.36) gives

$$S(E+) = \exp(2i\alpha), \tag{7.39}$$

where α is real, and

$$\alpha(E+) = P\tan^{-1}(q/b). \tag{7.40}$$

Notice that the phase shift α is an odd function of q, as is required.

We can use a finite product of factors like eqn (7.36) to give

$$S(E) = \prod_{j=1}^{N}\left(-\frac{q-ib_j}{q+ib_j}\right)^{P_j}, \tag{7.41}$$

where all b_j are positive and all P_j are real. On the physical cut eqn (7.41) gives

$$\alpha(E+) = \sum_{j=1}^{N} P_j\tan^{-1}(q/b_j). \tag{7.42}$$

Such functions can readily give S-wave resonances. For example, we choose $N = 2$ in eqn (7.41) and take

$$P_1 = -P_2 = 2, \quad \text{for } b_2 \gg b_1.$$

Near the threshold $\alpha \sim q$, so we have S-wave behaviour. For physical E, $0 \leqslant \alpha(E) < \pi$, and $\alpha(\infty) = 0$. There is a resonance at $E \simeq b_1^2$. The only singularities of $S(E)$ are a double pole at $E = -b_2^2$ on the physical sheet and a double pole at $E = -b_1^2$ on the second sheet.

A more general result

Suppose we choose a range $0 \leqslant E \leqslant B^2$. Then any continuous differentiable phase $\alpha(E)$ can be approximated as close as we like on this range by a series†

$$\sum_j P_j\tan^{-1}(q/b_j), \tag{7.43}$$

where all $b_j > B$. The series in eqn (7.43) corresponds to a function $S(E)$ whose only singularities on the physical or the second sheet lie on $-\infty \leqslant E \leqslant -B^2$.

It should, however, be noted that in general in order to improve the approximation at any stage, more terms having larger b_j must be added to eqn (7.43). This means that $S(E)$ gets more poles or branch points on the distant negative axis.

Lyng Petersen's theorem†

An even more striking result can be obtained by using conformal transformation techniques. The approximating function $S(E)$ can have

† J. Lyng Petersen (personal communication, 1970).

a phase which is arbitrarily close to the given (differentiable) phase $\alpha(E)$ on $0 \leqslant E \leqslant \infty$. In addition $S(E+)$ can be made to lie arbitrarily close to a given (differentiable) function $\sigma(E)$ on the left-hand cut $-\infty < E < E_1$, for $E_1 < 0$, *except* for a finite segment. The method of construction applies to all partial waves, and $S(E)$ has the proper threshold singularities at 0 and E_1. The function $S(E)$ and its continuation into the second sheet have no essential singularities and the function is finite, except on $-\infty < E < 0$ on some of the lower lying sheets.

APPENDIX I

USEFUL INTEGRALS AND LIMIT THEOREMS

Integrals

THE following simple integrals are useful for estimating limiting behaviour as $|s| \to \infty$ and for investigating similar problems about dispersion integrals.

For $0 < p < 1$,

$$\frac{1}{\pi}\int_0^\infty \frac{x^{p-1}}{1+x}\,dx = \frac{1}{\sin(\pi p)}.$$

Hence for $-\infty < s < 0$,

$$I(s,p) \equiv \frac{1}{\pi}\int_0^\infty \frac{(s')^{p-1}}{s'-s}\,ds' = \frac{(-s)^{p-1}}{\sin(\pi p)}, \tag{I.1}$$

with $(-s)^{p-1}$ defined to be real and positive on $-\infty < s < 0$. Also, for $-\infty < s < 0$,

$$I_L(s,p) \equiv \frac{1}{\pi}\int_0^\infty \frac{(s')^{p-1}\ln s'}{s'-s}\,ds'$$

$$= \frac{(-s)^{p-1}\ln(-s)}{\sin(\pi p)} - \pi\frac{\cos(\pi p)}{\sin^2(\pi p)}(-s)^{p-1}. \tag{I.2}$$

As an analytic function of s, $I(s,p)$ has the cut $0 \leqslant s \leqslant \infty$, and analytic continuation from eqn (I.1) gives

$$I(s\pm,p) = -\cot(\pi p)s^{p-1} \pm is^{p-1}, \quad \text{for } 0 < s < \infty. \tag{I.3}$$

The principal value integral is

$$PI(s,p) = -\cot(\pi p)s^{p-1}, \quad \text{for } 0 < s < \infty. \tag{I.4}$$

In particular for $p = \frac{1}{2}$,

$$I(s,\tfrac{1}{2}) = (-s)^{-\frac{1}{2}},$$

and for $0 < s < \infty$, $PI(s,\frac{1}{2}) = 0$.

Similarly $I_L(s,p)$ has the cut $0 \leqslant s \leqslant \infty$, and from eqn (I.2) analytic continuation gives, for $0 < s < \infty$,

$$I_L(s\pm,p) = -\cot(\pi p)s^{p-1}\ln s + \frac{\pi}{\sin^2(\pi p)}s^{p-1} \pm is^{p-1}\ln s. \tag{I.5}$$

The principal value integral is, for $0 < s < \infty$,

$$PI_L(s,p) = -\cot(\pi p)s^{p-1}\ln s + \frac{\pi}{\sin^2(\pi p)}\,s^{p-1}. \tag{I.6}$$

In particular $\qquad I_L(s,\tfrac{1}{2}) = (-s)^{-\frac{1}{2}}\ln(-s),$

and, for $0 < s < \infty$, $\qquad PI_L(s,\tfrac{1}{2}) = \pi s^{-\frac{1}{2}}.$

Finally, consider

$$J(s,0) = \frac{1}{\pi}\int_{s_0}^{\infty} \frac{ds'}{s'(s'-s)}, \quad \text{for } s_0 > 0.$$

For $-\infty < s < s_0$,

$$J(s,0) = \frac{1}{\pi(-s)}\ln(1-s/s_0) \tag{I.7}$$

and, for $s_0 < s < \infty$, the principal value integral is

$$PJ(s,0) = -\frac{1}{\pi s}\ln(s/s_0-1). \tag{I.8}$$

Limit theorems

LIMIT THEOREM A. *Let $\beta(s)$ be a real function defined on $1 \leqslant s \leqslant \infty$ and integrable on any finite range $1 \leqslant s \leqslant S$. Also let $\beta(s)$ obey the conditions*

(*a*) $\quad |\beta(s)-\beta(s')| < K|s-s'|^{\nu}$

for all s, s' on $[1,\infty]$, where K and ν are some positive constants,

(*b*) $\quad \lim_{s\to\infty} \beta(s)\ln s = 0,$

(*c*) $\quad \int_1^{\infty} \frac{\beta(s)\,ds}{s}$ *converges;*

then

$$\lim_{s\to\infty} P\int_1^{\infty} \frac{\beta(s')\,ds'}{s'-s} = 0.$$

Proof: Given any small positive number δ we can choose $\bar{s}$ such that

$$\left|\int_s^{\infty} \frac{\beta(s')\,ds'}{s'}\right| < \delta$$

and $\qquad |\beta(s)\ln s| < \delta$

for $s \geqslant \bar{s}$. We also choose $\bar{s} > 2$.

Now choose $s > 2\bar{s}$ and write

$$P\int_1^\infty \frac{\beta(s')\,ds'}{s'-s} = \left\{\int_1^{\bar{s}} + \int_{\bar{s}}^{\frac{1}{2}s} + \int_{\frac{1}{2}s}^{s-1/\bar{s}} + P\int_{s-1/\bar{s}}^{s+1/\bar{s}} + \int_{s+1/\bar{s}}^{2s} + \int_{2s}^{\infty}\right\}\frac{\beta(s')\,ds'}{s'-s}$$

$$= I_1+I_2+I_3+I_4+I_5+I_6,$$

where I_1 is the integral over the range $(1, \bar{s})$, I_2 the integral over the range $(\bar{s}, \frac{1}{2}s)$, and so on.

We have

$$|I_1| \leqslant \frac{1}{s-\bar{s}}\int_1^{\bar{s}} |\beta(s')|\,ds',$$

$$|I_2| \leqslant \delta\frac{\ln 2}{\ln \bar{s}} < \delta,$$

$$|I_3| < \frac{\delta}{\ln(\frac{1}{2}s)}\ln(\tfrac{1}{2}s\bar{s}) < 2\delta,$$

$$|I_5| < \frac{\delta}{\ln s}\ln(s\bar{s}) < 2\delta.$$

Also
$$I_4 = \int_{s-1/\bar{s}}^{s+1/\bar{s}} \frac{\beta(s')-\beta(s)}{s'-s}\,ds' + \beta(s)P\int_{s-1/\bar{s}}^{s+1/\bar{s}} \frac{ds'}{s'-s}. \tag{I.9}$$

The second integral on the right of eqn (I.9) vanishes, and using condition (*a*) of the theorem gives

$$|I_4| < \frac{2K}{\nu}(\bar{s})^{-\nu}.$$

We can write

$$I_6 = s\int_{2s}^\infty \frac{\beta(s')\,ds'}{s'(s'-s)} + \int_{2s}^\infty \frac{\beta(s')}{s'}\,ds',$$

so
$$|I_6| < \delta\frac{\ln 2}{\ln(2s)} + \delta < 2\delta.$$

Thus for $s > 2\bar{s}$ we have

$$\left|P\int_1^\infty \frac{\beta(s')\,ds'}{s'-s}\right| < \left(7\delta + \frac{2K}{\nu}(\bar{s})^{-\nu}\right) + \frac{1}{s-\bar{s}}\int_1^{\bar{s}} |\beta(s')|\,ds'. \tag{I.10}$$

We first choose δ so small and $\bar{s}$ so large that

$$\left(7\delta + \frac{2K}{\nu}(\bar{s})^{-\nu}\right) < \tfrac{1}{2}\epsilon,$$

where ϵ is as small as we like. Next we choose S so large that the second term on the right of eqn (I.10) is less than $\frac{1}{2}\epsilon$ for $s > S$. This proves the theorem.

Application of Theorem A

Suppose that $\alpha(s)$ obeys the Lipschitz condition (a) above (i.e. eqn (2.10) of § 2.1). Suppose further that the limit $\alpha(\infty)$ exists and that

$$\alpha(s) = \alpha(\infty)+\beta(s),$$

where $\beta(s)$ obeys conditions (b) and (c) above. We have

$$sP\int\limits_{s_0}^{\infty}\frac{\alpha(s')\,ds'}{s'(s'-s)} = \alpha(\infty)sP\int\limits_{s_0}^{\infty}\frac{ds'}{s'(s'-s)}+P\int\limits_{s_0}^{\infty}\frac{\beta(s')\,ds'}{s'-s}-\int\limits_{s_0}^{\infty}\frac{\beta(s')\,ds'}{s'}.$$

By eqn (I.8) and Theorem A, for large s the right-hand side equals

$$\alpha(\infty)\ln(s_0/s)-\int\limits_{s_0}^{\infty}\frac{\beta(s')}{s'}\,ds'+\gamma(s),$$

where $\gamma(s) \to 0$ as $s \to \infty$. Thus the function $\Lambda(s)$ defined in eqn (2.5) (§ 2.1) has the limiting behaviour

$$|\Lambda(s)| \underset{s\to+\infty}{\longrightarrow} Cs^{-2\alpha(\infty)/\pi}, \tag{I.11}$$

where
$$C = (s_0)^{2\alpha(\infty)/\pi}\exp\left(-\frac{2}{\pi}\int\limits_{s_0}^{\infty}\frac{\beta(s')}{s'}\,ds'\right).$$

It is easy to see that for any s in $s_0 \leqslant s \leqslant \infty$,

$$|\Lambda(s)| = C(s)s^{-2\alpha(\infty)/\pi},$$

where $0 < C_1 < C(s) < C_2$, C_1 and C_2 being constants. Moreover $C(s)$ will obey a Lipschitz condition on $[s_0, \infty]$ having the same index ν as in condition (a) above, as is shown in Theorem 106 on p. 145 of [33].

LIMIT THEOREM B. *The integral*

$$sP\int\limits_{2}^{\infty}\frac{ds'}{(s'-s)s'\ln s'}$$

has the value
$$-\ln\ln s+\ln\ln 2+O\left(\frac{1}{\ln s}\right),$$

as $s \to +\infty$.

Proof. We write

$$sP\int_2^\infty \frac{ds'}{(s'-s)s'\ln s'} = \int_2^{\frac{1}{2}s} \frac{ds'}{(s'-s)\ln s'} + P\int_{\frac{1}{2}s}^{2s} \frac{ds'}{(s'-s)\ln s'} - $$

$$- \int_2^{2s} \frac{ds'}{s'\ln s'} + s\int_{2s}^\infty \frac{ds'}{(s'-s)s'\ln s'}.$$

For $s > 80$ and $2 < s' < s$,

$$\frac{\ln s}{s^{\frac{1}{2}}} < \frac{\ln s'}{s'^{\frac{1}{2}}};$$

therefore for $s > 80$,

$$\left|\int_2^{\frac{1}{2}s} \frac{ds'}{(s'-s)\ln s'}\right| < \frac{s^{\frac{1}{2}}}{\ln s}\int_2^{\frac{1}{2}s} \frac{ds'}{s'^{\frac{1}{2}}(s-s')} < \frac{1}{\ln s}\int_0^{\frac{1}{2}} \frac{dx}{x^{\frac{1}{2}}(1-x)}.$$

Also

$$P\int_{\frac{1}{2}s}^{2s} \frac{ds'}{(s'-s)\ln s'} = P\int_{\frac{1}{2}}^{2} \frac{dx}{(\ln s+\ln x)(x-1)} = O\left(\frac{1}{\ln s}\right), \quad \text{as } s \to +\infty,$$

and

$$s\int_{2s}^\infty \frac{ds'}{s'(s'-s)\ln s'} = \int_2^\infty \frac{dx}{\ln(xs)x(x-1)} \leqslant \frac{\ln 2}{\ln 2s}.$$

Finally,

$$\int_2^{2s} \frac{ds'}{s'\ln s'} = \ln\ln 2s - \ln\ln 2$$

$$= \ln\ln s - \ln\ln 2 + O\left(\frac{1}{\ln s}\right), \quad \text{as } s \to +\infty.$$

The result follows directly.

Application of Theorem B

In § 2.7 we consider $\alpha(s)$ of the form

$$\alpha(s) = \alpha(\infty) + \frac{\pi H}{\ln s} + \beta(s), \quad \text{for } s_0 \leqslant s \leqslant \infty,$$

where $\alpha(\infty)$ and H are constants and the function $\beta(s)$ obeys the conditions (a), (b), and (c) in Theorem A. We also assume $s_0 > 1$.

Using eqn (I.8), Theorem A, and Theorem B we find

$$\frac{s}{\pi}P\int\limits_{s_0}^{\infty}\frac{\alpha(s')\,ds'}{s'(s'-s)} = -\frac{\alpha(\infty)}{\pi}\ln(s/s_0) - H\ln\ln s + H\ln\ln s_0 -$$

$$-\frac{1}{\pi}\int\limits_{s_0}^{\infty}\frac{\beta(s')}{s'}\,ds' + \gamma(s),$$

where

$$\lim_{s\to\infty}\gamma(s) = 0.$$

Therefore

$$\exp\left\{\frac{s}{\pi}P\int\limits_{s_0}^{\infty}\frac{\alpha(s')\,ds'}{s'(s'-s)}\right\} = C(s)\,\frac{1}{(\ln s)^H}\,s^{-\alpha(\infty)/\pi}, \quad \text{for } s_0 \leqslant s \leqslant \infty,$$

where

$$\lim_{s\to\infty}C(s) = s_0^{\,\alpha(\infty)/\pi}(\ln s_0)^H\exp\left\{-\frac{1}{\pi}\int\limits_{s_0}^{\infty}\frac{\beta(s')}{s'}\,ds'\right\},$$

and $C(s)$ obeys a Lipschitz condition with index ν.

APPENDIX II

MANIFOLDS OF SOLUTIONS OF A PARTIAL WAVE DISPERSION RELATION

No bound states

WE consider the case examined in § 2.3, that is the e.m.h. case with no bound states. It is convenient to use the notation of the elastic amplitude.

Let the solution $f(s)$ have the phase $\alpha(s)$ and the index $p \geqslant 1$. We shall show that there exists a positive number λ_1 such that the equation (eqn (2.43) of § 2.3)

$$\sin(\Delta\alpha(s,\lambda)) = \lambda q\xi(s)(s-s_0)^l|\Lambda(s)|\exp\left\{\frac{s}{\pi}P\int\limits_{s_0}^{\infty}\frac{\Delta\alpha(s',\lambda)}{s'(s'-s)}\,ds'\right\} \tag{II.1}$$

has a solution $\Delta\alpha(s,\lambda)$ $(s_0 \leqslant s \leqslant \infty)$, for $|\lambda| < \lambda_1$. $|\Lambda(s)|$ is given by eqn (2.7), and $\xi(s)$ is an arbitrary real polynomial of degree $(p-1)$, where the index p is the integer defined by

$$p < \frac{2\alpha(\infty)}{\pi} - l + \tfrac{1}{2} \leqslant p+1.$$

We only consider solutions $\Delta\alpha(s,\lambda)$ which obey the conditions (2.10) and (2.11) of § 2.1, and have $\Delta\alpha(s_0,\lambda) = 0$, $\Delta\alpha(\infty,\lambda) = 0$.

Let

$$h(s) = q\xi(s)(s-s_0)^l|\Lambda(s)|. \tag{II.2}$$

From the definition of p it follows that $h(\infty) = 0$, and that $h(s)$ obeys the condition (2.11). By Appendix I, on $s_0 \leqslant s \leqslant \infty$,

$$|\Lambda(s)| = C(s)s^{-2\alpha(\infty)/\pi},$$

where $0 < C_1 < C(s) < C_2$, C_1 and C_2 being constants. Also $C(s)$ obeys a Lipschitz condition, therefore $h(s)$ obeys a Lipschitz condition on $[s_0,\infty]$.

We can write eqn (II.1) in the form

$$\operatorname{im}\exp\left\{-\frac{s}{\pi}\int\limits_{s_0}^{\infty}\frac{\Delta\alpha(s',\lambda)\,ds'}{s'(s'-s)}\right\} = -\lambda h(s), \quad \text{for } s_0 \leqslant s \leqslant \infty. \tag{II.3}$$

Therefore

$$\exp\left\{-\frac{s}{\pi}\int\limits_{s_0}^{\infty}\frac{\Delta\alpha(s',\lambda)\,ds'}{s'(s'-s)}\right\} = -\frac{\lambda}{\pi}\int\limits_{s_0}^{\infty}\frac{h(s')\,ds'}{s'-s} + Q(s), \tag{II.4}$$

where $Q(s)$ is a real polynomial. Equation (II.4) is true because the function on the left-hand side has no singularities other than the cut $s_0 \leqslant s \leqslant \infty$, and the discontinuity there is given by eqn (II.3). By Limit Theorem A of Appendix I,

$$\frac{s}{\pi}\int_{s_0}^{\infty} \frac{\Delta\alpha(s',\lambda)}{s'(s'-s)}\,ds' \to -\frac{1}{\pi}\int_{s_0}^{\infty} \frac{\Delta\alpha(s',\lambda)}{s'}\,ds'$$

as $s \to +\infty$, and

$$\int_{s_0}^{\infty} \frac{h(s')\,ds'}{s'-s} \to 0$$

as $s \to +\infty$. Therefore $Q(s)$ is a positive constant Q. Putting $s = 0$ in eqn (II.4) we find

$$Q = 1+\frac{\lambda}{\pi}\int_{s_0}^{\infty} \frac{h(s')}{s'}\,ds'. \tag{II.5}$$

Thus eqn (II.4) becomes

$$\exp\left\{-\frac{s}{\pi}\int_{s_0}^{\infty} \frac{\Delta\alpha(s',\lambda)\,ds'}{s'(s'-s)}\right\} = G(s,\lambda), \tag{II.6}$$

where

$$G(s,\lambda) = 1-\frac{\lambda s}{\pi}\int_{s_0}^{\infty} \frac{h(s')\,ds'}{s'(s'-s)}. \tag{II.7}$$

Our problem then is to show that there exists a number λ_1, such that we can solve eqn (II.6) for $\Delta\alpha(s,\lambda)$ for $|\lambda| < \lambda_1$. We can only have a solution if $G(s,\lambda) \neq 0$ for all s (real and complex). For $|s| \to \infty$, $G(s,\lambda) \to Q$; also $G(0,\lambda) = 1$, and $G(s,0) = 1$. Clearly there exists a positive number λ_1 such that

$$Q > 0 \quad \text{for } |\lambda| < \lambda_1,$$

and such that for all s,

$$G(s,\lambda) \neq 0, \quad \text{for } |\lambda| < \lambda_1.$$

For $|\lambda| < \lambda_1$ the only singularity of the real analytic function $\ln G(s,\lambda)$ is the cut $s_0 \leqslant s \leqslant \infty$. Also for $|\lambda| < \lambda_1$, $|\ln G(s,\lambda)|$ is bounded as $|s| \to \infty$. Since $h(s)$ has at most a finite number of changes of sign on $[s_0, \infty]$, it follows that

$$\arg G(s+,\lambda) \to N\pi \quad \text{as } s \to +\infty,$$

where N is an integer (positive, negative, or zero). Thus we can write the dispersion relation

$$\ln G(s,\lambda) = \frac{s}{\pi}\int_{s_0}^{\infty} \frac{\arg G(s'+,\lambda)\,ds'}{s'(s'-s)}. \tag{II.8}$$

Equations (II.8) and (I.7) show that

$$\ln G(s,\lambda) \sim \ln(-s), \quad \text{as } s \to -\infty,$$

unless $N = 0$. Since $\ln G(s,\lambda) \to \ln Q$ as $s \to -\infty$, we must have $N = 0$, i.e.

$$\arg G(s+,\lambda) \to 0, \quad \text{as } s \to +\infty.$$

For $|\lambda| < \lambda_1$, a solution of eqn (II.6) is

$$\Delta\alpha(s,\lambda) = -\arg G(s+,\lambda), \quad \text{for } s_0 \leqslant s \leqslant \infty,$$

$$= -\arg\left\{1-\frac{\lambda s}{\pi}\int\limits_{s_0}^{\infty}\frac{h(s')\,ds'}{s'(s'-s)}\right\}. \tag{II.9}$$

This is a solution to eqn (II.1) and it is contained in the set $\mathscr{A}$ (§ 2.3). We can write eqn (II.9) as

$$\tan(\Delta\alpha(s,\lambda)) = \lambda h(s)\Big/\left\{1-\frac{\lambda s}{\pi}P\int\limits_{s_0}^{\infty}\frac{h(s')\,ds'}{s'(s'-s)}\right\}. \tag{II.10}$$

It follows from eqn (II.9) that $\Delta\alpha(\infty,\lambda) = 0$.

Thus for $|\lambda| < \lambda_1$, $\alpha(s)+\Delta\alpha(s,\lambda)$ are the phase shifts of other partial wave solutions lying in the same connected region of $\mathscr{A}$ as the solution $f(s)$ whose phase is $\alpha(s)$. Since there are p arbitrary parameters in $\xi(s)$, these solutions form a p-dimensional manifold.

Other solutions

Above we only found solutions in a convex connected region of $\mathscr{A}$ around $f(s)$. Consider now eqn (2.41) of § 2.3. Its general solution can be found similarly. Let

$$H(s) = qE(s)(s-s_0)^l|\tilde{\Lambda}_1(s)|,$$

where $E(s)$ is any real polynomial. The solution is

$$\tan(\Delta\alpha(s)) = H(s)\Big/\left\{Q(s)-\frac{1}{\pi}\prod_{j=1}^{M}(s-s_j)P\int\limits_{s_0}^{\infty}\frac{H(s')\,ds'}{\prod\limits_{j=1}^{M}(s'-s_j)(s'-s)}\right\},$$

where M is a suitably chosen positive integer and the real numbers s_j $(j = 1,2,\ldots,M)$ are on $-\infty < s < s_0$. Also $Q(s)$ is a real polynomial.

Several bound states

We sketch the proof of Theorem 3 of § 2.4. We consider solutions of a given p.w.d.r. in the e.m.h. case having phase shifts in $\mathscr{A}$. For simplicity we discuss the elastic case.

Let a solution $f(s)$ have the phase $\alpha(s)$ and n bound states at B_i $(i = 1,2,\ldots,n)$, having residues Γ_i $(i = 1,2,\ldots,n)$. The index p of $f(s)$ is the integer obeying

$$p < 2\left(\frac{\alpha(\infty)}{\pi}+n\right)-l+\tfrac{1}{2} \leqslant p+1.$$

Let $f'(s)$ be a distinct solution with phase $(\alpha(s)+\Delta\alpha(s))$, and m bound states at B'_i $(i = 1, 2,..., m)$. Then

$$(f'(s)-f(s)) \prod_{i=1}^{n} (s-B_i) \prod_{i=1}^{m} (s-B'_i)$$
$$= E(s)(s-s_0)^l \Lambda(s) \exp\left\{\frac{s}{\pi} \int_{s_0}^{\infty} \frac{\Delta\alpha(s')\,ds'}{s'(s'-s)}\right\}, \tag{II.11}$$

where $\Lambda(s)$ is given by eqn (2.5) and $E(s)$ is a polynomial of degree $(r-1)$.

The norm of eqn (II.11) gives

$$\sin(\Delta\alpha(s)) = \psi(s) \prod_{i=1}^{m} (s-B'_i)^{-1} \exp\left\{\frac{s}{\pi} P \int_{s_0}^{\infty} \frac{\Delta\alpha(s')\,ds'}{s'(s'-s)}\right\}, \quad \text{for } s_0 \leqslant s \leqslant \infty, \tag{II.12}$$

with

$$\psi(s) = qE(s)(s-s_0)^l |\Lambda(s)| \prod_{i=1}^{n} (s-B_i)^{-1}, \quad \text{for } s_0 \leqslant s \leqslant \infty. \tag{II.13}$$

Solutions $f'(s)$ lying in the same connected region of $\mathscr{A}$ as $f(s)$ must have $\Delta\alpha(\infty) = 0$. Equation (II.12) shows that this requires

$$2\left(\frac{\alpha(\infty)}{\pi}+n\right)-l+\tfrac{1}{2} > r-(m-n). \tag{II.14}$$

This condition can only be satisfied for m such that $p+m-n > 0$. Then $p+m-n$ is the largest value of r for which eqn (II.14) is obeyed.

Thus if $\Delta\alpha(\infty) = 0$ (and $\Delta\alpha(s) \in \mathscr{A}$) we have $0 < r \leqslant p+m-n$. Also $\psi(s) \prod_{i=1}^{m} (s-B'_i)^{-1}$ belongs to $\mathscr{A}$, so by the results of the section above dealing with no bound states, $\Delta\alpha(s)$ is given by

$$\exp\left\{-\frac{s}{\pi} \int_{s_0}^{\infty} \frac{\Delta\alpha(s')}{s'(s'-s)}\,ds'\right\} = 1-\frac{s}{\pi} \int_{s_0}^{\infty} \frac{\psi(s') \prod_{i=1}^{m} (s'-B'_i)^{-1}}{s'(s'-s)}\,ds' \tag{II.15}$$

for all s, and

$$\tan(\Delta\alpha(s)) = \psi(s) \prod_{i=1}^{m} (s-B'_i)^{-1} \left\{1-\frac{s}{\pi} P \int_{s_0}^{\infty} \frac{\psi(s') \prod_{i=1}^{m} (s'-B'_i)^{-1}}{s'(s'-s)}\,ds'\right\}^{-1} \tag{II.16}$$

for $s_0 \leqslant s \leqslant \infty$.

Letting $s \to B_j$ $(j = 1, 2,..., n)$ in eqn (II.11) gives

$$\prod_{i=1}^{m} (B_j-B'_i) = C_j, \quad \text{for } j = 1, 2,..., n, \tag{II.17}$$

where $$C_j = K_j \exp\left\{\frac{B_j}{\pi}\int\limits_{s_0}^{\infty} \frac{\Delta\alpha(s')}{s'(s'-B_j)}\,ds'\right\}, \tag{II.18}$$

and $$K_j = -E(B_j)(B_j-s_0)^l \Lambda(B_j)\Gamma_j^{-1} \prod_{\substack{i=1\\(i\neq j)}}^{n} (B_j-B_i)^{-1}. \tag{II.19}$$

If now $\Delta\alpha(\infty) = 0$, eqn (II.15) gives

$$C_j = K_j\left\{1-\frac{B_j}{\pi}\int\limits_{s_0}^{\infty} \frac{\psi(s')\prod\limits_{i=1}^{m}(s'-B_i')^{-1}}{s'(s'-B_j)}\,ds'\right\}^{-1}, \tag{II.20}$$

so C_j depends on B_i' $(i = 1, 2,..., m)$.

If $m < n$, eqns (II.17), (II.18), (II.19), and (II.20) cannot be satisfied for any set of values B_i' $(i = 1, 2,..., m)$ in the limit where $\Delta\alpha(s)$ (and hence also $E(s)$ and C_j) goes to zero. Thus there is a neighbourhood of $f(s)$ which can only contain solutions with n or more bound states. This proves part (*c*) of Theorem 3.

Let m be an integer such that $m \geqslant n$ and $p+m-n > 0$, and let $\xi(s)$ be an arbitrary polynomial of degree $(p+m-n-1)$. It can be shown that there exists a λ' $(\lambda' > 0)$ such that, replacing $E(s)$ by $\lambda\xi(s)$, eqns (II.17), (II.18), (II,19), and (II.20) have an $(m-n)$-dimensional manifold $\mathscr{M}$ of solutions B_i' $(i = 1, 2,..., m)$ for $|\lambda| < \lambda'$. Let $\bar{B} = (\bar{B}_{n+1},..., \bar{B}_m)$ be $(m-n)$ arbitrary distinct real numbers, all less than s_0 and different from B_j $(j = 1, 2,..., n)$. Inserting $B_i' = \bar{B}_i$ $(i = n+1,..., m)$ in eqns (II.17), (II.18), (II.19), and (II.20) we have n equations for the remaining B_i' $(i = 1, 2,..., n)$. For $\lambda = 0$ a solution is $B_i' = B_i$ $(i = 1, 2,..., n)$. Schauder's theorem (see [34]) shows that there is a $\bar{\lambda}$ $(\bar{\lambda} > 0)$ such that eqns (II.17), (II.18), (II.19), and (II.20) have solutions

$$B_i' = B_i(\lambda, \bar{B}) \quad (i = 1, 2,..., n) \qquad \text{for } |\lambda| < \bar{\lambda}.$$

We can choose $\bar{\lambda}$ such that for fixed λ $(|\lambda| < \bar{\lambda})$ the $B_i(\lambda, \bar{B})$ are distinct real numbers less than $s_0-\epsilon$ where $\epsilon > 0$. Also there is an $(m-n)$-dimensional manifold of vectors $\bar{B}$ such that $\lambda' > 0$ where $\lambda' = \min(\bar{\lambda}(\bar{B}))$.

There exists a λ_1 $(0 < \lambda_1 \leqslant \lambda')$ such that the right-hand side of eqn (II.15) does not vanish for any s (real or complex) when $|\lambda| < \lambda_1$ and B_i' $(i = 1, 2,..., m)$ is a solution in $\mathscr{M}$. Equation (II.15) then gives a solution of eqns (II.12) and (II.13) when $|\lambda| < \lambda_1$, and B_i' $(i = 1, 2,..., m)$ is a solution in $\mathscr{M}$.

We can now readily deduce parts (*b*) and (*d*) of Theorem 3, and part (*a*) follows by using Lemma 4.

APPENDIX III

SOLUTIONS OF THE INTEGRAL EQUATIONS

THE integral equations used in Chapters 4 and 5 (eqns (4.36) and (5.40)) are of the form

$$D(z,\lambda) = 1+\lambda z^2 \int\limits_{p_1}^{p_2} K(z,z')z'\bar{\rho}(z')D(z',\lambda)\,\mathrm{d}z', \qquad \text{(III.1)}$$

where $K(z,z')$ is real, positive, and symmetric for $p_1 \leqslant z \leqslant p_2$, $p_1 \leqslant z' \leqslant p_2$. We shall assume that $\bar{\rho}(z)$ is non-negative on $[p_1, p_2]$ and is $L_2(p_1, p_2)$. We can then prove Theorem C.

THEOREM C. *There exists a positive constant λ_1 such that the solution $D(z,\lambda)$ of eqn* (III.1) *is a monotonically increasing function of λ in $0 \leqslant \lambda < \lambda_1$, for each fixed z in $p_1 \leqslant z \leqslant p_2$. Further, for $\lambda \simeq \lambda_1$ and $p_1 \leqslant z \leqslant p_2$,*

$$D(z,\lambda) = \frac{\lambda_1^2\, C(z)}{\lambda_1-\lambda}+O(1), \qquad \text{(III.2)}$$

where $C(z)$ is positive and bounded in $p_1 \leqslant z \leqslant p_2$. Therefore for $p_1 \leqslant z \leqslant p_2$,

$$D(z,\lambda) \to +\infty, \qquad \text{as } \lambda \to \lambda_1 \quad (\lambda < \lambda_1).$$

The constant λ_1 obeys

$$\lambda_1 \geqslant \left\{ \int\limits_{p_1}^{p_2}\int\limits_{p_1}^{p_2} \mathrm{d}z\mathrm{d}z'\; z^3\bar{\rho}(z)\big(K(z,z')\big)^2 z'^3\bar{\rho}(z') \right\}^{-\frac{1}{2}}. \qquad \text{(III.3)}$$

We rely on fairly well-known properties of Fredholm type equations. Let $\mathscr{M}$ be the set of points in $[p_1, p_2]$ where $\bar{\rho}(z) > 0$. When $z \in \mathscr{M}$ we can write eqn (III.1) as

$$d(z,\lambda) = \psi(z)+\lambda \int\limits_{\mathscr{M}} \bar{K}(z,z')d(z',\lambda)\,\mathrm{d}z' \quad (z \in \mathscr{M}), \qquad \text{(III.4)}$$

where

$$\psi(z) = (\bar{\rho}(z)/z)^{\frac{1}{2}},$$

$$d(z,\lambda) = \psi(z)D(z,\lambda),$$

and

$$\bar{K}(z,z') = \big(z^3\bar{\rho}(z)\big)^{\frac{1}{2}}K(z,z')\big(z'^3\bar{\rho}(z')\big)^{\frac{1}{2}}.$$

Equation (III.4) is of the Fredholm type and has a real, positive, and symmetric kernel. The solution of eqn (III.1) is given by

$$D(z,\lambda) = 1+\lambda z^2 \int\limits_{\mathscr{M}} K(z,z')\big(\bar{\rho}(z')z'^3\big)^{\frac{1}{2}}d(z',\lambda)\,\mathrm{d}z'. \qquad \text{(III.5)}$$

Let the integral operator T be defined on the set of all functions f in $L_2(\mathscr{M})$ by

$$(Tf)(z) = \int_{\mathscr{M}} \bar{K}(z,z')f(z')\,dz'. \tag{III.6}$$

Now T is a completely continuous Hermitian operator (see [35], pp. 147–8). Such an operator has a pure point spectrum of real eigenvalues. The set of eigenvalues $\{\kappa_i\}$ can only have one accumulation point $\kappa = 0$, so all non-zero eigenvalues are discrete (see [35], p. 173).

Since $\bar{K}(z,z') > 0$ on $\mathscr{M}\times\mathscr{M}$ there exists (see [36], § 13.7) a positive eigenvalue κ_1 such that all other eigenvalues κ_i obey $|\kappa_i| < \kappa_1$. This eigenvalue κ_1 is non-degenerate and the corresponding eigenfunction is a multiple of a function $\psi_1(z)$ which is positive almost everywhere on $\mathscr{M}$. Also

$$\kappa_1 \leqslant \Big(\int_{p_1}^{p_2}\int_{p_1}^{p_2} (\bar{K}(z,z'))^2\,dzdz'\Big)^{\frac{1}{2}}.$$

If λ^{-1} is not an eigenvalue of T, eqn (III.4) has a unique solution

$$d(z,\lambda) = \frac{a\psi_1(z)}{1-\lambda\kappa_1} + \tilde{d}(z,\lambda), \tag{III.7}$$

where

$$\tilde{d}(z,\lambda) = \sum_{\kappa_i\neq\kappa_1} \frac{(P_i\psi)(z)}{1-\lambda\kappa_i}. \tag{III.8}$$

Here a is the positive constant

$$a = \int_{\mathscr{M}} \psi(z)\psi_1(z)\,dz\Big/\int_{\mathscr{M}} (\psi_1(z))^2\,dz.$$

In eqn (III.8), $(P_i\psi)(z)$ is the projection of $\psi(z)$ on to the space of eigenfunctions of T having the eigenvalue κ_i. The sum in eqn (III.8) converges in the mean and $\tilde{d}(z,\lambda)$ is defined for almost all z in $\mathscr{M}$. Also

$$\int_{\mathscr{M}} (\tilde{d}(z,\lambda))^2\,dz \leqslant \delta^{-2}\int_{\mathscr{M}} (\psi(z))^2\,dz, \tag{III.9}$$

where

$$\delta = \min_{\kappa_i\neq\kappa_1} |1-\lambda\kappa_i|.$$

Equations (III.5), (III.7), and (III.8) give

$$D(z,\lambda) = 1 + \frac{\lambda C(z)}{1-\lambda\kappa_1} + \tilde{D}(z,\lambda), \tag{III.10}$$

where

$$C(z) = az^2\int_{\mathscr{M}} K(z,z')(\bar{\rho}(z')z'^3)^{\frac{1}{2}}\psi_1(z')\,dz'$$

and

$$\tilde{D}(z,\lambda) = \lambda z^2\int_{\mathscr{M}} K(z,z')(\bar{\rho}(z')z'^3)^{\frac{1}{2}}\tilde{d}(z',\lambda)\,dz'.$$

Clearly $C(z)$ is positive and bounded for $p_1 \leqslant z \leqslant p_2$.

By Schwarz's inequality and eqn (III.9),

$$|\tilde{D}(z,\lambda)| \leqslant \lambda\delta^{-1}|z|^2\left\{\int\limits_{p_1}^{p_2} |K(z,z')|^2\bar{\rho}(z')z'^3\,dz'\right\}^{\frac{1}{2}}\left\{\int\limits_{\mathscr{M}} (\psi(z'))^2\,dz'\right\}^{\frac{1}{2}}.$$

Thus $\tilde{D}(z,\lambda)$ is bounded on $p_1 \leqslant z \leqslant p_2$ if λ^{-1} is not an eigenvalue κ_i $(\kappa_i \neq \kappa_1)$ of T. Writing $\lambda_1 = \kappa_1^{-1}$ we have eqns (III.2) and (III.3).

Since $\lambda_1 < |\kappa_i|^{-1}$, for $i \neq 1$, it follows (see [35], p. 30) that if $0 < \lambda < \lambda_1$, eqn (III.4) can be solved by iteration. Thus

$$d(z,\lambda) = \psi(z)+\lambda(T\psi)(z)+\lambda^2(T^2\psi)(z)+\dots \qquad \text{(III.11)}$$

The sum converges in the mean in $\mathscr{M}$. The terms on the right of eqn (III.11) are all positive, so if $0 \leqslant \lambda' < \lambda'' < \lambda_1$, then $0 < d(z,\lambda') < d(z,\lambda'')$ for almost all z in $\mathscr{M}$. By eqn (III.5),

$$D(z,\lambda') < D(z,\lambda''), \quad \text{for } p_1 \leqslant z \leqslant p_2.$$

This completes the proof of Theorem C.

Total positivity

It can be shown in the e.m.h. case that the kernel (eqn (4.36))

$$K(z,z') = \frac{1}{4\pi}\frac{1}{z+z'} \quad (z,z' \in [p_1,p_2]),$$

is strictly totally positive (s.t.p.). Then $\bar{K}(z,z')$ is also s.t.p., and the operator T in eqn (III.6) has some interesting properties. There is a theorem (definitions and theorems on total positivity are discussed in [37], pp. 11, 35, and 149) which states that the operator T has an enumerable set of eigenvalues $\kappa_1 > \kappa_2 > \kappa_3 > \dots$ which are positive and non-degenerate. The only limit point of the set $\{\kappa_i\}$ is zero. Let $\phi_i(z)$ be the corresponding eigenfunctions of T; then $\phi_i(z)$ has $(i-1)$ nodal zeros in $[p_1, p_2]$. Moreover the zeros of $\phi_i(z)$ separate those of $\phi_{i+1}(z)$.

The eigenvalues are $\kappa_i = 1/\lambda_i$, where $\lambda = \lambda_i$ when the ith bound state is at s_0. For $\lambda \simeq \lambda_i$,

$$D(z,\lambda) \simeq \frac{c}{\lambda-\lambda_i}\phi_i(z).$$

It follows that for $\lambda \gtrsim \lambda_2$, $D(z,\lambda)$ has one bound state zero near $z = 0$ and another in $[p_1,p_2]$. For $\lambda \gtrsim \lambda_3$, $D(z,\lambda)$ has a bound state zero near $z = 0$ and two in $[p_1,p_2]$, and so on. It is clear that the bound state zeros can never move into the region $p_2 < z < \infty$. Thus there is a limit on the strength of the binding which is not dependent on the strength of the interaction.

We have not proved that the kernel in the π–π case (eqn (5.40)) is s.t.p., so we do not know whether such results hold in that case.

APPENDIX IV

TURNING POINTS OF im $D = 0$ FOR π–π SCATTERING

HERE the results quoted at the end of § 5.5 are derived. A *turning point* was defined as a point where the tangent to im $D = 0$ in the ϕ-plane is parallel to the v-axis.

We need only consider the region $\mathscr{P}$: $-\frac{1}{2}\pi \leqslant u \leqslant 0,\ 0 \leqslant v \leqslant \infty$. Because of the singularity at $z = -1$ in $I(z, z')$ (see eqn (5.44)) it is convenient to use

$$\tilde{I}(z,z') = \frac{\cos 2u + \cosh 2v}{\sinh 2v} I(z,z'), \quad \text{for } p_1 \leqslant z' \leqslant p_2,$$

$$= \frac{2|z^2-1|}{\sinh 2v} I(z,z').$$

By eqn (5.43) the locus im $D = 0$ is given by

$$\int\limits_{p_1}^{p_2} \tilde{I}(z,z')\sigma(z')\,dz' = 0, \tag{IV.1}$$

so at a turning point,

$$\int\limits_{p_1}^{p_2} \partial_v \tilde{I}(z,z')\sigma(z')\,dz' = 0. \tag{IV.2}$$

We define (analogous to eqn (5.61))

$$\tilde{J}(z;z',z'') = \tilde{I}(z,z')\partial_v \tilde{I}(z,z'') - \tilde{I}(z,z'')\partial_v \tilde{I}(z,z')$$

$$= \left(\frac{\cos 2u + \cosh 2v}{\sinh 2v}\right)^2 J(z;z',z'').$$

At each turning point

$$\int\limits_{p_1}^{p_2} \tilde{J}(z;p_1,z')\sigma(z')\,dz' = 0 \tag{IV.3}$$

and

$$\int\limits_{p_1}^{p_2} \tilde{J}(z;z',p_2)\sigma(z')\,dz' = 0. \tag{IV.4}$$

We have to examine the branch structure and signs of $\partial_v \tilde{I}(z,p) = 0$ and $\tilde{J}(z;z',z'') = 0$ to find the region of $\mathscr{P}$ where eqns (IV.3) and (IV.4) can be satisfied.

Equations (5.44) and (5.30) show that if $1 \leqslant p < \infty$,

$$\tilde{I}(z,p) = \pi \frac{\cosh^2 v}{\cosh^2 v - p^2}, \quad \text{for } u = -\tfrac{1}{2}\pi, \tag{IV.5}$$

and

$$\tilde{I}(z,p) = \tfrac{1}{2}\pi \frac{\sinh^2 v}{\sinh^2 v + p^2}, \quad \text{for } u = 0. \tag{IV.6}$$

Therefore

$$\partial_v \tilde{I}(z,p) < 0, \quad \text{for } u = -\tfrac{1}{2}\pi,$$
$$\partial_v \tilde{I}(z,p) > 0, \quad \text{for } u = 0.$$

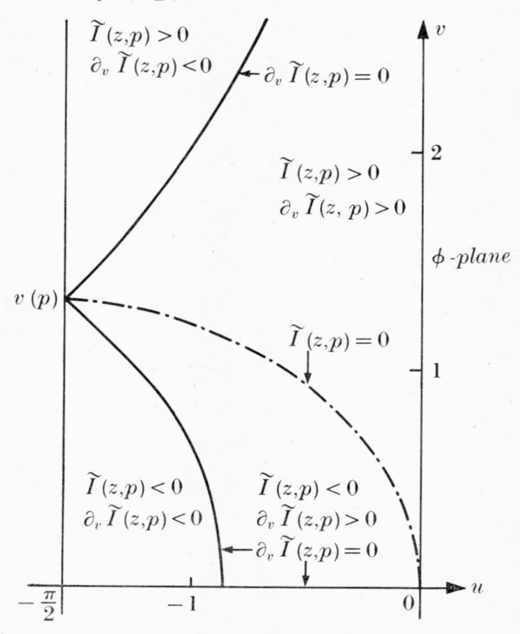

FIG. IV.1. The curves $\tilde{I}(z,p) = 0$, $\partial_v \tilde{I}(z,p) = 0$ for $p = 2$, and the signs of $\tilde{I}(z,p)$, $\partial_v \tilde{I}(z,p)$.

Equations (5.28), (5.33), and (5.44) show that

$$\left.\begin{array}{l} \tilde{I}(z,p) \text{ is finite} \\ \partial_v \tilde{I}(z,p) = 0 \end{array}\right\} \quad \text{for } v = 0. \tag{IV.7}$$

Since $\tilde{I} > 0$ above the locus $\tilde{I} = 0$ in $\mathscr{P}$, we have $\partial_v \tilde{I} > 0$ on $\tilde{I} = 0$ in $\mathscr{P}$. These results are shown in Fig. IV.1.

We shall write $v(p) = \operatorname{arc\,cosh} p$. For $p > 1$ two branches of $\partial_v \tilde{I}(z,p) = 0$ leave $u = -\frac{1}{2}\pi$ at $v(p)$ with slopes ± 1. (For $p = 1$ there are three branches leaving $v(p) = 0$ with slopes $0, \pm\sqrt{7}$.)

Ignoring the line $v = 0$, there are only two branches of $\partial_v \tilde{I}(z,p) = 0$ in $\mathscr{P}$; that with the positive (negative) slope will be called the upper (lower) branch (see Fig. IV.1). The upper branch is asymptotic to $u = 0$.

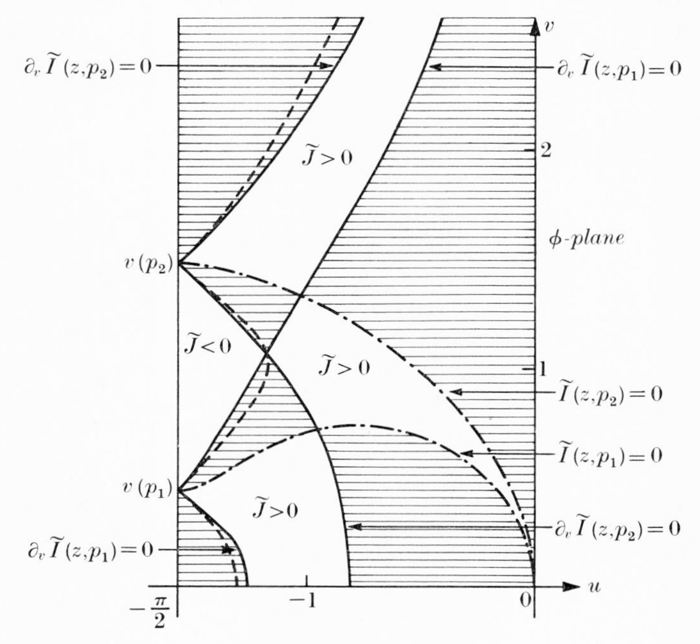

FIG. IV.2. The curves are:

– · – · – · – · – $\tilde{I}(z, p_i) = 0 \quad (i = 1, 2)$;

———————— $\partial_v \tilde{I}(z, p_i) = 0 \quad (i = 1, 2)$;

– – – – – – – – $\tilde{J}(z; p_1, p_2) = 0$.

The regions (unshaded) where $\tilde{J}$ has definite sign are also shown. The figure is drawn for $p_1 = 1{\cdot}1$, $p_2 = 2{\cdot}3$.

We shall assume that if $z'' > z'$, the upper (lower) branch of $\partial_v \tilde{I}(z, z'') = 0$ lies entirely above the upper (lower) branch of $\partial_v \tilde{I}(z, z') = 0$. This is true in a variety of examples which have been examined.

By eqns (IV.5) and (IV.6),

$$\tilde{J}(z; z', z'') = (\tilde{I}(z, z'))^2 \frac{\partial}{\partial v}\left(\frac{\sinh^2 v + z'^2}{\sinh^2 v + z''^2}\right) > 0$$

for $u = 0$ and $1 \leqslant z' < z''$. Also

$$\tilde{J}(z; z', z'') = (\tilde{I}(z, z'))^2 \frac{\partial}{\partial v}\left(\frac{\cosh^2 v - z'^2}{\cosh^2 v - z''^2}\right) < 0$$

for $u = -\frac{1}{2}\pi$ and $1 \leqslant z' < z''$.

By eqn (IV.7),

$$\tilde{J}(z; z', z'') = 0, \quad \text{for } v = 0.$$

Also if $1 \leqslant z' < z''$, $\tilde{J}(z; z', z'') > 0$ on $\tilde{I}(z, z') = 0$ and on $\tilde{I}(z, z'') = 0$.

Using Fig. IV.1 we find that $\tilde{J}(z; p_1, p_2)$ is of definite sign in the four unshaded regions in Fig. IV.2. The locus $\tilde{J}(z; p_1, p_2) = 0$ must lie in the shaded regions. A similar result holds for $\tilde{J}(z; z', z'') = 0$ where $1 \leqslant z' < z''$.

At $\phi = (-\frac{1}{2}\pi, v(p))$, $\partial_v \tilde{I}(z,p)$ has a singularity of higher order than $\tilde{I}(z,p)$. Therefore two branches of $\tilde{J}(z; p_1, p_2) = 0$ leave $u = -\frac{1}{2}\pi$ at $v(p_1)$ and at $v(p_2)$. They are tangential to the branches of $\partial_v \tilde{I}(z,p_1) = 0$ and $\partial_v \tilde{I}(z,p_2) = 0$ respectively. In Fig. IV.2 the dashed line shows $\tilde{J}(z; p_1, p_2) = 0$. A similar result holds for $\tilde{J}(z; z', z'') = 0$, for $1 \leqslant z' < z''$.

Thus $\tilde{J}(z; z', z'') = 0$ has three branches. We call them the first, second, and third branches, in order of increasing v. The second branch joins $v(z')$ to $v(z'')$ on $u = -\frac{1}{2}\pi$, and the third branch is asymptotic to $u = 0$.

Let $E(p_1, p_2)$ be the region covered by the families of second branches of $\tilde{J}(z; p_1, z') = 0$ and $\tilde{J}(z; z', p_2) = 0$ for $p_1 \leqslant z' \leqslant p_2$. As in § 5.5, $\bar{u}(p_1, p_2)$ denotes the maximum value of u on the second branch of $\tilde{J}(z; p_1, p_2) = 0$. Values of $\bar{u}(p_1, p_2)$ are given in Fig. 5.12.

Now we have our main result. For fixed p_1, $\bar{u}(p_1, p_2)$ increases as p_2 increases, and for fixed p_2 it decreases as p_1 increases. Hence the maximum value of u on $E(p_1, p_2)$ is $\bar{u}(p_1, p_2)$.

We should mention an unimportant complication. If z obeys

$$\tilde{J}(z; p_1, z') = 0, \quad \tilde{J}(z; p_1, p_2) = 0,$$

then

$$\tilde{J}(z; z', p_2) = 0.$$

The third branch of $\tilde{J}(z; p_1, z') = 0$ meets the second branch of $\tilde{J}(z; p_1, p_2) = 0$. At the intersection, the second branch of $\tilde{J}(z; z', p_2) = 0$ crosses the second branch of $\tilde{J}(z; p_1, p_2) = 0$. Similarly the second branch of $\tilde{J}(z; p_1, z') = 0$ must cross the second branch of $\tilde{J}(z; p_1, p_2) = 0$. Hence $E(p_1, p_2)$ is somewhat larger than the region in $\mathscr{P}$ bounded by the second branch of $\tilde{J}(z; p_1, p_2) = 0$. It remains that $\bar{u}(p_1, p_2)$ is the maximum value of u on $E(p_1, p_2)$.

It is easy to show that the turning points of $\operatorname{im} D = 0$ can only lie where

$$\tilde{I}(z, p_1) \,.\, \tilde{I}(z, p_2) < 0.$$

Also, if $p_1 \leqslant z' \leqslant p_2$, then $\tilde{J}(z; p_1, z')$ is positive in the region outside $E(p_1, p_2)$ where $\partial_v \tilde{I}(z, p_1) > 0$. Equation (IV.3) cannot be satisfied for any z in this region. Similarly eqn (IV.4) cannot be satisfied in the region outside $E(p_1, p_2)$ where $\partial_v \tilde{I}(z, p_2) > 0$. Thus the turning points of $\operatorname{im} D = 0$ must lie in $E(p_1, p_2)$.

Given z_0 in $E(p_1, p_2)$ then $\tilde{J}(z_0; p_1, z') = 0$ or $\tilde{J}(z_0; z', p_2) = 0$ for some z' $(p_1 \leqslant z' \leqslant p_2)$. Therefore as mentioned in § 5.5 we can construct a function $\bar{\rho}(z)$, for $p_1 \leqslant z \leqslant p_2$, such that $\operatorname{im} D = 0$ has a branch with a turning point at z_0. Hence $E(p_1, p_2)$ is the smallest region containing turning points for the general $\bar{\rho}(z)$ $(\bar{\rho} \geqslant 0)$.

REFERENCES

1. L. Castillejo, R. H. Dalitz, and F. J. Dyson, *Phys. Rev.* **101,** 453 (1956).
2. G. Frye and R. L. Warnock, *Phys. Rev.* **130,** 478 (1963).
3. R. L. Warnock, *Phys. Rev.* **131,** 1320 (1963).
4. D. H. Lyth, *J. math. Phys.* **11,** 2646 (1970).
5. M. Froissart, *Nuovo Cim.* **22,** 191 (1961).
6. C. Lovelace, *Commun. math. Phys.* **4,** 261 (1967).
7. F. G. Tricomi, *Integral Equations* (Interscience, N.Y. 1957).
8. J. Hamilton and W. S. Woolcock, *Rev. mod. Phys.* **35,** 737 (1963).
9. N. Levinson, *Mat.-fys. Meddr* **25,** Nr. 9 (1949).
10. A. Martin, *Nuovo Cim.* **15,** 99 (1960).
11. R. G. Newton, *J. math. Phys.* **1,** 319 (1960).
12. J. M. Jauch, *Helv. phys. Acta* **30,** 143 (1957).
13. J. Hamilton, *The Dynamics of Elementary Particles and the Pion–Nucleon Interaction* (*Course B*) (Nordita lecture notes, Nordita 1968).
14. D. Atkinson, K. Dietz, and D. Morgan, *Ann. Phys.* **37,** 77 (1966).
15. D. Atkinson and M. B. Halpern, *Phys. Rev.* **150,** 1377 (1966).
16. E. C. Titchmarsh, *The Theory of Functions* (Clarendon Press, Oxford 1939).
17. T. Kinoshita, *Phys. Rev. Lett.* **16,** 869 (1966); *Phys. Rev.* **154,** 1438 (1967).
18. G. H. Hardy and J. E. Littlewood, *Proc. Lond. math. Soc.* **30,** 23 (1930).
19. D. Atkinson and D. Morgan, *Nuovo Cim.* **41,** 559 (1966).
20. R. E. Peierls, *1954 Conference on Nuclear and Meson Physics, Glasgow* (Pergamon, London 1955).
21. H. Burkhardt, *Dispersion Relation Dynamics* (North-Holland, Amsterdam 1969).
22. I. J. R. Aitchison and P. R. Graves-Morris, *Nucl. Phys.* **B14,** 683 (1969).
23. G. Höhler, *Z. Phys.* **152,** 546 (1958).
24. A. Donnachie and J. Hamilton, *Ann. Phys.* **31,** 410 (1965).
25. G. F. Chew and S. Mandelstam, *Phys. Rev.* **119,** 467 (1960).
26. J. Lyng Petersen, *Nucl. Phys.* **B13,** 73 (1969).
27. A. Donnachie and J. Hamilton, *Phys. Rev.* **138,** B678 (1965).
28. G. F. Chew and F. E. Low, *Phys. Rev.* **101,** 1570 (1956).
29. J. Hamilton, 'Pion–Nucleon Interactions' in *High Energy Physics, Vol. I,* Ed. E. Burhop (Academic Press, N.Y. 1967).
30. G. Calucci, L. Fonda, and G. C. Ghirardi, *Phys. Rev.* **166,** 1719 (1968).
31. G. Calucci and G. C. Ghirardi, *Phys. Rev.* **169,** 1339 (1968).
32. E. Hille, *Analytic Function Theory, Vol. II* (Ginn & Co., Boston 1962).
33. E. C. Titchmarsh, *The Theory of Fourier Integrals* (2nd edition, Clarendon Press, Oxford 1948).

34. M. A. Krasnosel'skii, *Topological Methods in the Theory of Nonlinear Integral Equations* (Pergamon Press, London 1964).

35. L. A. Ljusternik and W. I. Sobolew, *Elemente der Funktionalanalysis* (Akademie Verlag, Berlin 1960).

36. A. C. Zaanen, *Linear Analysis* (North-Holland, Amsterdam 1953).

37. S. Karlin, *Total Positivity, Vol. I* (Stanford University Press, 1968).

AUTHOR INDEX

SUBJECT INDEX

Date Due

			UML 735